1

INTRODUCTION TO VECTOR CALCULUS

CLASSIFICATION OF VECTORS

Broadly vectors can be classified into two categories–

(i) **Axial Vectors :** Where a vector has rotational motion lying along the normal to the plane of rotation of the body and remains unchanged under inversion. e.g.: Torque, angular momentum etc.

(ii) **Polar Vector :** Where a vector has linear motion in a particular direction but changes under inversion or reflection. e.g. displacement, position vector, velocity etc.

Some special vectors

(i) **Unit vector :** It is a vector with unit magnitude and characterizes the direction of the vector mathematically it is denoted by

$$\hat{A} = \frac{\vec{A}}{|A|}$$

In Cartesian coordinate system, let us choose three unit vectors along three mutually perpendicular axes as $\hat{i}, \hat{j}$ and $\hat{k}$ in x, y and z directions respectively. Then any arbitrary vector $\vec{A}$ can be expressed as

$$\vec{A} = A_x \hat{i} + A_y \hat{j} + A_z \hat{k}$$

where A_x, A_y and A_z are called the components of $\vec{A}$ in x, y and z directions.

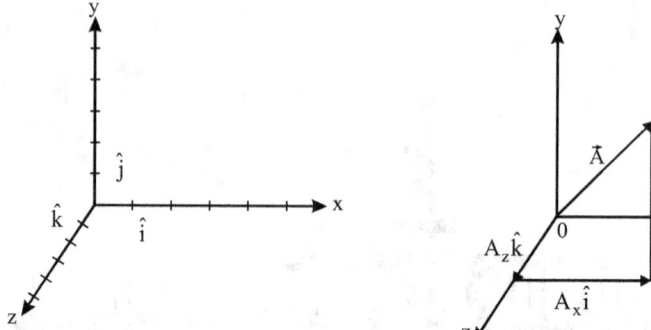

The magnitude of $\vec{A}$ is given using parallelogram law

$$\left|\vec{A}\right| = \sqrt{A_x^2 + A_y^2 + A_z^2}$$

hence unit vector along $\vec{A}$ is given by

$$\hat{A} = \frac{A_x\hat{i} + A_y\hat{j} + A_z\hat{k}}{\sqrt{A_x^2 + A_y^2 + A_z^2}}$$

Direction cosines

The cosines of the angles, which $\vec{A}$ makes with x, y and z axis are called direction cosines of the vector. If l, m and n are the direction cosines along $\overrightarrow{ox}$, $\overrightarrow{oy}$ and $\overrightarrow{oz}$ axes, then

$$l = \cos\alpha = \frac{A_x}{\left|\vec{A}\right|} \quad \text{or} \quad A_x = l\left|\vec{A}\right|$$

$$m = \cos\beta = \frac{A_y}{\left|\vec{A}\right|} \quad \text{or} \quad A_y = m\left|\vec{A}\right|$$

$$n = \cos\gamma = \frac{A_z}{\left|\vec{A}\right|} \quad \text{or} \quad A_z = n\left|\vec{A}\right|$$

and

$$l^2 + m^2 + n^2 = \frac{A_x^2 + A_y^2 + A_z^2}{\left|\vec{A}\right|^2} = \frac{\left|\vec{A}\right|^2}{\left|\vec{A}\right|^2} = 1$$

so

$$l^2 + m^2 + n^2 = \cos^2\alpha + \cos^2\beta + \cos^2\gamma = 1$$

and

$$\vec{A} = \left|\vec{A}\right|\left(l\hat{i} + m\hat{j} + n\hat{k}\right)$$

and the unit vector

$$\hat{a} = \frac{\vec{A}}{\left|\vec{A}\right|} = l\hat{i} + m\hat{j} + n\hat{k}$$

(ii)   **Null Vector :** Any vector with magnitude zero is called null vector. It is collinear with every vector and denoted by $\vec{O}$.

(iii)   **Collinear or parallel vector :** When vectors are parallel, then these are collinear vectors, whatsoever their magnitudes may be. Direction of there vectors may be some or opposite.

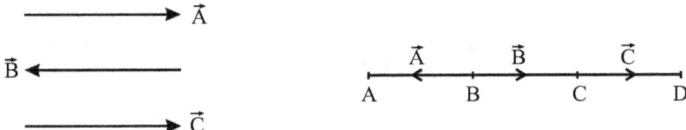

When any scalar is multiplied to any vector then the resultant vector becomes collinear with original one. e.g. $\vec{B} = \lambda\vec{A}$ i.e. $\vec{B}$ vector is λ times $\vec{A}$ with same direction as of $\vec{A}$.

(iv)   **Coplanar vectors :** When vectors lies in the same geometrical plane they are called coplanar vectors. Otherwise these are called non-coplanar vectors.

(v)   **Like vectors :** The collinear vectors with same sense of direction irrespective of magnitude are called like vectors.

(vi)   **Reciprocal vectors :** When the magnitude of a vector is reciprocal to the magnitude of other vector with same direction then it is called reciprocal vector. It is written as $\frac{1}{A}$ i.e. $\vec{A}^{-1} = \frac{\hat{a}}{\left|\vec{A}\right|}$ where $\hat{a}$ is the unit vector along the direction of $\vec{A}$.

Product of vectors

(i)   **Scalar product or dot product :**

When the result of product of two vectors is a scalar quantity then this product is known as scalar (or dot) product of the given vectors.

Mathematically it is obtained by multiplying the magnitudes of the vectors with cosines of the angle between them

i.e. $\qquad \vec{A}.\vec{B} = AB\cos\theta$

$$\vec{A}.\vec{B} = |\vec{A}||\vec{B}|\cos\theta$$

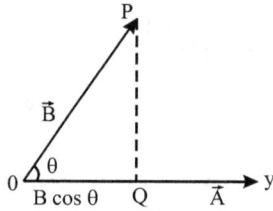

Alternatively scalar product may be defined as multiplication of one vector with component of another in the direction of first.

In case of Cartesian unit vectors

$$\hat{i}.\hat{i} = i.i\cos 0 = 1.1 = 1 = \hat{j}.\hat{j} = \hat{k}.\hat{k}$$

and $\qquad \hat{i}.\hat{j} = ij\cos 90° = 0 = \hat{j}.\hat{k} = \hat{i}.\hat{k}$

if $\qquad \vec{A} = A_x\,\hat{i} + A_y\,\hat{j} + A_z\,\hat{k}$

and $\qquad \vec{B} = B_x\,\hat{i} + B_y\,\hat{j} + B_z\,\hat{k}$

then $\qquad \vec{A}.\vec{B} = A_x B_x + A_y B_y + A_z B_z$

(a) Scalar product obeys commutative law i.e. $\vec{A}.\vec{B} = \vec{B}.\vec{A}$

(b) It obeys distributive law i.e. $\vec{A}.(\vec{B}+\vec{C}) = \vec{A}.\vec{B} + \vec{A}.\vec{C}$

(c) Two non-zero vectors are orthogonal or perpendicular when $\theta = 90°$ i.e. $\cos\theta = 0$

then $\qquad \vec{A}.\vec{B} = 0$

similarly two vectors are collinear when $\theta = 0$ or π i.e. $\cos\theta = \pm 1$

then for $\theta = 0$, $\qquad \vec{A}.\vec{B} = AB$

and for $\theta = \pi$, $\qquad \vec{A}.\vec{B} = -AB$

Physical examples–

(i) Work done $W = \vec{F}.\vec{ds}$

(ii)   Power $= \vec{F}.\vec{v}$

(iii)   Magnetic flux of a magnetic field $= \vec{B}.\overrightarrow{ds}$ where $\vec{B}$ is magnetic flux density over an area $\overrightarrow{ds}$.

(iv)   Electric flux of an electric field $= \vec{E}.\overrightarrow{ds}$ where $\vec{E}$ the electric field intensity through elementary area $\overrightarrow{ds}$.

(ii)   Vector product or cross product

When the product of two vectors is a vector quantity, then the product is called vector product or cross product mathematically it is written as

$$\vec{C} = \vec{A} \times \vec{B}$$

$$= \left|\vec{A}\right|\left|\vec{B}\right| \sin \theta.\,\hat{n} \quad \text{where } 0 \le \theta \le \pi$$

here $\hat{n}$ is the unit vector in the direction of normal to the plane containing $\vec{A}$ and $\vec{B}$ such that $\vec{A}, \vec{B}$ and $\vec{C}$ from a right handed coordinate system with rotation from $\vec{A}$ to $\vec{B}$.

for $\hat{i}, \hat{j}$ and $\hat{k}$.

$$\hat{i} \times \hat{i} = i\,i \sin 0 = 0 = \hat{j} \times \hat{j} = \hat{k} \times \hat{k}$$

and

$$\hat{i} \times \hat{j} = i\,j \sin 90 = 1 = \hat{j} \times \hat{k} = \hat{k} \times \hat{i}$$

but

$$\hat{j} \times \hat{i} = j\,i \sin(-90) = -1 = \hat{k} \times \hat{j} = \hat{i} \times \hat{k}$$

if

$$\vec{A} = A_x\hat{i} + A_y\,\hat{j} + A_z\hat{k}$$

and

$$\vec{B} = B_x\hat{i} + B_y\,\hat{j} + B_z\hat{k}$$

then

$$\vec{A} \times \vec{B} = \left|\vec{A}\right|\left|\vec{B}\right| \sin \theta\,\hat{n}$$

$$= \begin{vmatrix} \hat{i} & \hat{j} & \hat{k} \\ A_x & A_y & A_z \\ B_x & B_y & B_z \end{vmatrix}$$

or

$$\vec{A} \times \vec{B} = \left(A_yB_z - A_zB_y\right)\hat{i} - \left(A_xB_z - A_zB_x\right)\hat{j} + \left(A_xB_y - A_yB_x\right)$$

and
$$\hat{n} = \frac{\vec{A} \times \vec{B}}{\left|\vec{A} \times \vec{B}\right|}$$

hence
$$\sin \theta = \frac{\left|\vec{A} \times \vec{B}\right|}{\left|\vec{A}\right|\left|\vec{B}\right|}$$

where
$$\left|\vec{A}\right| = \sqrt{A_x^2 + A_y^2 + A_z^2}$$

and
$$\left|\vec{B}\right| = \sqrt{B_x^2 + B_y^2 + B_z^2}$$

if the rotation from $\vec{A}$ to $\vec{B}$ is anti clockwise then $\vec{C} = \vec{A} \times \vec{B}$ is +ve. and if rotation is clockwise then $\vec{C} = \vec{A} \times \vec{B}$ is –ve.

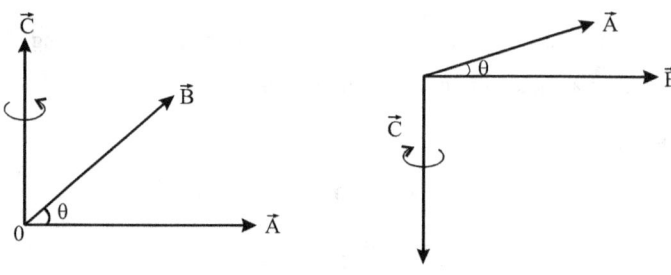

Properties

(a) Cross product is not commutative

$$\vec{A} \times \vec{B} \neq -\vec{B} \times \vec{A} \text{ but } \vec{A} \times \vec{B} = -\vec{B} \times \vec{A}$$

(b) It is distributive i.e. $\vec{A} \times (\vec{B} + \vec{C}) = \vec{A} \times \vec{B} + \vec{A} \times \vec{C}$

(c) If two vectors are collinear or parallel then

$$\theta = 0 \text{ or } \pi$$

then
$$\sin \theta = 0$$

and
$$\vec{A} \times \vec{B} = 0$$

(d) Two vectors are perpendicular then $\theta = 90°$

so
$$\vec{A} \times \vec{B} = \left|\vec{A}\right|\left|\vec{B}\right|\hat{n}$$

Some examples

(i) Moment of forces

$$\vec{\tau} = \vec{r} \times \vec{F}$$

(ii) Angular momentum $\vec{L} = \vec{r} \times \vec{p} = m(\vec{r} \times \vec{v})$

(iii) Linear velocity $\vec{v} = \vec{\omega} \times \vec{r}$

(iv) force on a charged particle

$$\vec{F} = q(\vec{v} \times \vec{B}) \text{ where q is in coulombs.}$$

(v) Force on a charged particle moving through electric and magnetic field is

$$\vec{F} = q(\vec{E} + \vec{v} \times \vec{B}). \text{ this is known as Lorentz force.}$$

Scalar Triple product

When a vector is scalarly multiplied with the cross product of other two vectors then the result is called scalar triple product.

$$\left[\vec{A}\,\vec{B}\,\vec{C}\right] = \vec{A}.(\vec{B} \times \vec{C}) = \vec{B}.(\vec{C} \times \vec{A}) = \vec{C}.(\vec{A} \times \vec{B})$$

$$\vec{A}.(\vec{B} \times \vec{C}) = \begin{vmatrix} A_x & A_y & A_z \\ B_x & B_y & B_z \\ C_x & C_y & C_z \end{vmatrix}$$

Vector triple product

When any vector is vectorily multiplied with vector product of other two vectors taken in cyclic order then the result is known as vector triple product.

$$\vec{A} \times (\vec{B} \times \vec{C}) = \vec{B}(\vec{A}.\vec{C}) - \vec{C}(\vec{A}.\vec{B})$$

$$\vec{B} \times (\vec{C} \times \vec{A}) = \vec{C}(\vec{B}.\vec{A}) - \vec{A}(\vec{B}.\vec{C})$$

$$\vec{C} \times (\vec{A} \times \vec{B}) = \vec{A}(\vec{C}.\vec{B}) - \vec{B}(\vec{C}.\vec{A})$$

Properties

(i) $\vec{A} \times (\vec{B} \times \vec{C}) + \vec{B} \times (\vec{C} \times \vec{A}) + \vec{C} \times (\vec{A} \times \vec{B}) = 0$

(ii) $(\vec{A} \times \vec{B}) \times \vec{C} = -\vec{C} \times (\vec{A} \times \vec{B}) = -\vec{A}(\vec{B}.\vec{C}) + \vec{B}(\vec{A}.\vec{C})$

(iii) $\left(\vec{A} \times \vec{B}\right).\left(\vec{C} \times \vec{D}\right) = \left(\vec{A}.\vec{C}\right)\left(\vec{B}.\vec{D}\right) - \left(\vec{A}.\vec{D}\right)\left(\vec{B}.\vec{C}\right)$

(iv) $\vec{A} \times \left(\vec{B} \times \left(\vec{C} \times \vec{D}\right)\right) = \vec{B}\left(\vec{A}.\left(\vec{C} \times \vec{D}\right)\right) - \left(\vec{A}.\vec{B}\right)\left(\vec{C} \times \vec{D}\right)$

Vector differentiation

This is the limiting value of ratio of a vector to the change of a scalar as the change tends to zero is called vector differentiation.

$$\frac{d\vec{f}}{du} = \lim_{\delta u \to 0} \frac{\delta f}{\delta u}$$

$$= \lim_{\delta u \to 0} \frac{\vec{\delta}\left(u + \delta u\right) - \vec{f}\left(u\right)}{\delta u}$$

Properties

(a) $\dfrac{d}{du}\left(\vec{A} \pm \vec{B}\right) = \dfrac{d\vec{A}}{du} \pm \dfrac{d\vec{B}}{du}$

(b) $\dfrac{d}{du}\left(\vec{A}.\vec{B}\right) = \dfrac{d\vec{A}}{du}.\vec{B} + \vec{A}.\dfrac{d\vec{B}}{du}$

(c) $\dfrac{d}{du}\left(\vec{A} \times \vec{B}\right) = \vec{A} \times \dfrac{d\vec{B}}{du} + \dfrac{d\vec{A}}{du} \times \vec{B}$

(d) $\dfrac{d}{du}\left(\vec{A}.\left(\vec{B} \times \vec{C}\right)\right) = \vec{A}.\left(\vec{B} \times \dfrac{d\vec{C}}{du}\right) + \vec{A}.\left(\dfrac{d\vec{B}}{du} \times \vec{C}\right) + \dfrac{d\vec{A}}{du}.\left(\vec{B} \times \vec{C}\right)$

(e) $\dfrac{d}{du}\left\{\vec{A} \times \left(\vec{B} \times \vec{C}\right)\right\} = \vec{A} \times \left(\vec{B} \times \dfrac{d\vec{C}}{du}\right) + \vec{A} \times \left(\dfrac{d\vec{B}}{du} \times \vec{C}\right) + \dfrac{d\vec{A}}{du} \times \left(\vec{B} \times \vec{C}\right)$

(f) $\dfrac{d}{dt}\vec{a}(s) = \dfrac{d\vec{a}}{ds}\dfrac{ds}{dt}$

(g) $\dfrac{d}{dt}(\vec{e}.\vec{e}) = \vec{e}.\dfrac{d\vec{e}}{dt} + \dfrac{d\vec{e}}{dt} \cdot \vec{e} = 2\vec{e} \cdot \dfrac{d\vec{e}}{dt}$

or $\dfrac{d}{dt}\left(\vec{e}^2\right) = 2\vec{e}.\dfrac{d\vec{e}}{dt}$

(h) $\dfrac{d}{dt}\left\{\phi\,\vec{a}(t)\right\} = \phi\dfrac{d\vec{a}}{dt}$

Fields

A field is a region in space, where a physical quantity can be specified at every point of the region. Fields may be classified as either scalar or vector depending upon the type of function involved i.e. if a scalar function is taken care of, the field is called a scalar field or if a vector function is involved then that is called a vector field. The temperature of the atmosphere, the height of the surface of earth above sea level are examples of scalar fields. The wind velocity, the gravity force on a mass in space or the force on a charged body in an electric field are examples of vectors fields.

Vector differential operator

The vector differential operator (del operator) is given by

$$\vec{\nabla} = \dfrac{\partial}{\partial x}\,\hat{i} + \dfrac{\partial}{\partial y}\,\hat{j} + \dfrac{\partial}{\partial z}\,\hat{k}$$

It remains invariant under rotation of coordinate system.

Directional Derivatives

Suppose $f(x, y, z)$ and $f\left(x + \Delta x, y + \Delta y, z + \Delta z\right)$ are two scalar point functions and

$$df = \dfrac{\partial f}{\partial x}\,dx + \dfrac{\partial f}{\partial y}\,dy + \dfrac{\partial f}{\partial z}\,dz$$

Variation of f can be written as

$$\dfrac{df}{dr} = \dfrac{\partial f}{\partial x}\dfrac{dx}{dr} + \dfrac{\partial f}{\partial y}\dfrac{dy}{dr} + \dfrac{\partial f}{\partial z}\dfrac{dz}{dr}$$

$$= \left(\dfrac{\partial f}{\partial x}\hat{i} + \dfrac{\partial f}{\partial y}\hat{j} + \dfrac{\partial f}{\partial z}\hat{k}\right).\left(\dfrac{dx}{dr}\hat{i} + \dfrac{dy}{dr}\hat{j} + \dfrac{dz}{dr}\hat{k}\right)$$

$$\dfrac{df}{dr} = \vec{\nabla}f.\hat{b}$$

Where unit vector $\qquad \hat{b} = \dfrac{dx}{dr}\hat{i} + \dfrac{dy}{dr}\hat{j} + \dfrac{dz}{dr}\hat{k}$

Hence the directional derivative is

$$\boxed{\frac{df}{dr} = \vec{\nabla}f \cdot \frac{\vec{a}}{|\vec{a}|}}$$ where f can be a vector function also .

Coordinates systems

(i)   Cartesian coordinates system

Any point P(x, y, z) can be represented as below–

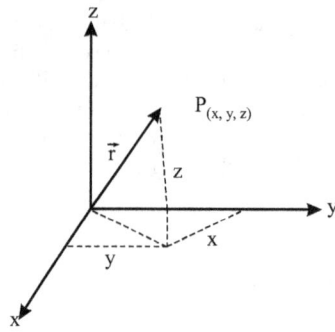
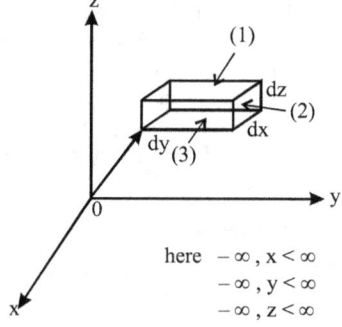

here $-\infty, x < \infty$
$-\infty, y < \infty$
$-\infty, z < \infty$

The position vector for P can be

$$\vec{r} = x\hat{i} + y\hat{j} + z\hat{k}$$

and increase in $\vec{r}$

$$d\vec{r} = dx\,\hat{i} + dy\,\hat{j} + dz\,\hat{k}$$

corresponding volume element

$$dV = dx\,dy\,dz.$$

Surface areas of different sections are

$$ds_1 = dy\,dx\,\hat{k}$$

$$ds_2 = dx\,dz\,\hat{j}$$

$$ds_3 = dy\,dz\,\hat{i}$$

and surface area of areas opposite to these surfaces

$$ds_1' = -dy\,dx\,\hat{k}$$

$$ds_2' = -dx\,dz\,\hat{j}$$

$$ds_3' = -dy\,dz\,\hat{i}$$

(ii) Cylindrical coordinate system

Here any point is represented in term of cylindrical coordinate (ρ, ϕ, z) as shown.

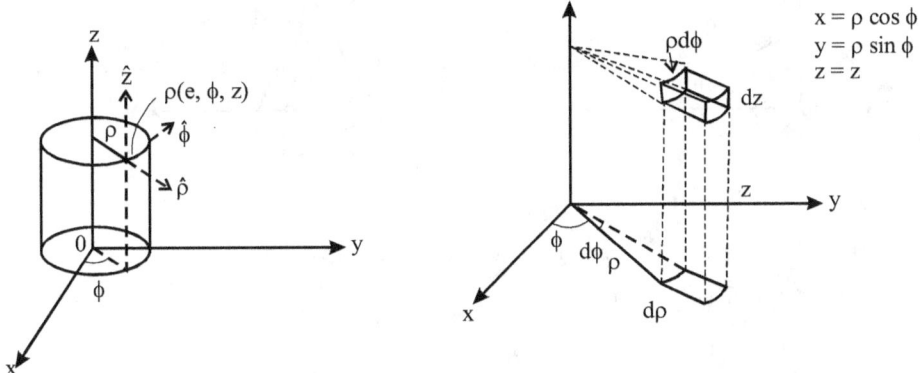

Any vector in cylindrical coordinate system can be written as

$$\vec{A} = A_\rho \,\hat{\rho} + A_\phi \hat{\phi} + A_z \hat{z}$$

differential length i.e. increase in length is given by

$$d\vec{l} = d\rho \,\hat{\rho} + \rho \, d\phi \,\hat{\phi} + dz \,\hat{z}$$

and

$$dl = \sqrt{d\rho^2 + (\rho d\phi)^2 + (dz)^2}$$

differential surfaces are

$$ds_1 = \rho \, d\phi \, dz \,\hat{\rho}$$

$$ds_2 = dz \, d\rho \,\hat{\phi}$$

$$ds_3 = \rho \, d\phi \, d\rho \,\hat{z}$$

Differential volume is

$$dV = d\rho \,\rho d\phi \, dz$$

$$= \rho \, d\rho \, d\phi \, dz$$

(iii) Spherical coordinates system

In this system any point at the surface of a sphere can be represented in terms of spherical coordinates (r, θ, ϕ) as shown.

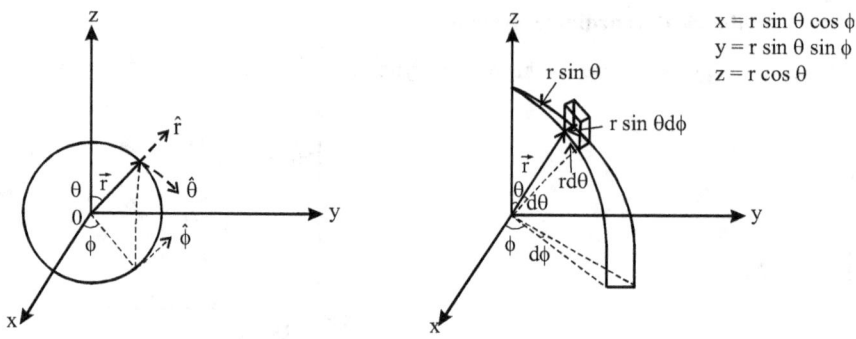

Here r is the radius of imaginary sphere with $0 \leq r \leq \infty$.

$0 \leq \theta \leq \pi$ and $0 \leq \phi \leq 2\pi$.

Here differential length $\overline{dl} = A_r \hat{r} + A_\theta \hat{\theta} + A_\phi \hat{\phi}$

$$\overline{dl} = dr\,\hat{r} + rd\theta\,\hat{\theta} + r\sin\theta d\phi\,\hat{\phi}$$

differential surfaces are

$$ds_1 = rd\theta\,\hat{\theta} \times r\sin\theta d\phi\,\hat{\phi} = r^2\sin\theta\,d\theta\,d\phi\,\hat{r}$$

$$ds_2 = r\sin\theta d\phi\,\hat{\phi} \times dr\,\hat{r} = r\sin\theta\,d\phi\,dr\,\hat{\theta}$$

$$ds_3 = dr\,\hat{r} \times rd\theta\,\hat{\theta} = rdr\,d\theta\,\hat{\phi}$$

and differential volume is

$$dV = dr(rd\theta)(r\sin\theta\,d\phi)$$

$$= r^2\sin\theta\,d\theta\,dr\,d\phi$$

The Gradient of a scalar point function

When del operator $(\vec{\nabla})$ operates on a scalar point function f(x, y, z), the obtained vector is called gradient $(\vec{\nabla}f)$ of a scalar.

$$\vec{\nabla}f = \text{grad } f = \left(\frac{\partial}{\partial x}\hat{i} + \frac{\partial}{\partial y}\hat{j} + \frac{\partial}{\partial z}\hat{k}\right)f(x,y,z)$$

so, $$\vec{\nabla}f = \frac{\partial f}{\partial x}\hat{i} + \frac{\partial f}{\partial y}\hat{j} + \frac{\partial f}{\partial z}\hat{k}$$

Gradient in Cylindrical Coordinate System

here $$\psi(r) = \psi(\rho, \phi, z)$$ ...(1)

so $$d\psi = \frac{\partial \psi}{\partial \rho} d\rho + \frac{\partial \psi}{\partial \phi} d\phi + \frac{\partial \psi}{\partial z} dz$$ ...(2)

and $$d\psi = d\vec{r} \text{ grad } \psi$$ ...(3)

But the line element in cylindrical coordinates system

$$d\vec{r} = d\rho \,\hat{\rho} + \rho d\phi \,\hat{\phi} + dz \,\hat{z}$$ ...(4)

and any vector

$$\vec{A}(r) = \vec{A}(\rho, \phi, z) = A_\rho \,\hat{\rho} + A_\phi \,\hat{\phi} + A_z \,\hat{z}$$ ...(5)

so $$\text{grad } \psi = (\text{grad } \psi)_\rho \,\hat{\rho} + (\text{grad } \psi)_\phi \,\hat{\phi} + (\text{grad } \psi)_z \,\hat{z}$$ ...(6)

so $$d\psi = d\vec{r} . \text{ grad } \psi$$

$$= (\text{grad } \psi)_\rho \, d\rho + (\text{grad } \psi)_\phi \, \rho d\phi + (\text{grad } \psi)_z \, dz$$ ...(7)

comparing (2) and (7)

$$\frac{\partial \psi}{\partial \rho} = (\text{grad } \psi)_\rho \,;\, \frac{\partial \psi}{\partial \phi} = \rho(\text{grad } \psi)_\phi \,;\, \frac{\partial \psi}{\partial z} = (\text{grad } \psi)_z$$

so from (6)

$$\text{grad } \psi = \vec{\nabla}\psi = \frac{\partial \psi}{\partial \rho} \,\hat{\rho} + \frac{1}{\rho}\frac{\partial \psi}{\partial \phi} \,\hat{\phi} + \frac{\partial \psi}{\partial z} \,\hat{z}$$

So del operator in cylindrical system is

$$\vec{\nabla} \equiv \frac{\partial}{\partial \rho} \,\hat{\rho} + \frac{1}{\rho}\frac{\partial}{\partial \phi} \,\hat{\phi} + \frac{\partial}{\partial z} \,\hat{z}$$

Gradient in spherical coordinate system

here $$\psi(r) = \psi(r, \theta, \phi)$$

$$d\psi = \frac{\partial \psi}{\partial r} dr + \frac{\partial \psi}{\partial \theta} d\theta + \frac{\partial \psi}{\partial \phi} d\phi$$ ...(1)

again $$d\psi = d\vec{r} . \text{ grad } \psi$$ ...(2)

line element in spherical system

$$d\vec{r} = dr\,\hat{r} + r d\theta\,\hat{\theta} + r\sin\theta\,d\phi\,\hat{\phi} \qquad \qquad ...(3)$$

Any vector in spherical system

$$\vec{A}(r) = \vec{A}(r, \theta, \phi) = A_r\,\hat{r} + A_\theta\,\hat{\theta} + A_\phi\,\hat{\phi} \qquad ...(4)$$

$$\text{grad }\psi = (\text{grad }\psi)_r\,\hat{r} + (\text{grad }\psi)_\theta\,\hat{\theta} + (\text{grad }\phi)_\phi\,\hat{\phi} \qquad ...(5)$$

as
$$d\psi = d\vec{r}\cdot\text{grad }\psi$$

$$= (\text{grad }\psi)_r\,dr + (\text{grad }\psi)_\theta\,rd\theta + (\text{grad }\psi)_\phi\,r\sin\theta\,d\phi\,...(6)$$

comparing equation (1) and (6)

$$(\text{grad }\psi)_r = \frac{\partial\psi}{\partial r};\ (\text{grad }\psi)_\theta = \frac{1}{r}\frac{\partial\psi}{\partial\phi}$$

and
$$(\text{grad }\psi)_\phi = \frac{1}{r\sin\theta}\frac{\partial\psi}{\partial\phi}$$

so
$$\text{grad }\psi = \vec{\nabla}\psi = \frac{\partial\psi}{\partial r}\,\hat{r} + \frac{1}{r}\frac{\partial\psi}{\partial\theta}\,\hat{\theta} + \frac{1}{r\sin\theta}\frac{\partial\psi}{\partial\phi} \qquad ...(7)$$

so del operator in spherical coordinates

$$\vec{\nabla} \equiv \frac{\partial}{\partial r}\hat{r} + \frac{1}{r}\frac{\partial}{\partial\theta}\hat{\theta} + \frac{1}{r\sin\theta}\frac{\partial}{\partial\phi}$$

Properties of Gradient

(i) $\nabla(u+v) = \nabla v + \nabla u$

(ii) $\nabla(vu) = v\nabla u + u\nabla v$

(iii) $\nabla\left(\dfrac{u}{v}\right) = \dfrac{v\nabla u - u\nabla v}{v^2}$

(iv) $\nabla v^n = nv^{n-1}\,\nabla v$

(v)   if $\vec{A} = \nabla T$, T is scalar potential of $\vec{A}$.

(vi)   Line integrals of gradients are path independent.

(vii)   If $\nabla X = 0$, it refers to a maximum or minimum point of the function X.

(viii)   A vector field derived from the gradient of a scalar field is called as Lameller field.

(ix)   Grad V or $\vec{\nabla} V$ points in the direction of the maximum rate of change in V.

(x)   the projection of $\left(\vec{\nabla} V\right)$ in the direction of a unit vector is $\vec{\nabla} V . \vec{a}$ and called directional derivative of V along $\vec{a}$.

Divergence of a vector (at a point on the vector field)

When del operator is operated as dot product on a differentiable vector field, the obtained scalar is called the divergence of the vector.

$$\text{div } \vec{A} = \vec{\nabla} . \vec{A} = \left(\frac{\partial}{\partial x} \hat{i} + \frac{\partial}{\partial y} \hat{j} + \frac{\partial}{\partial z} \hat{k} \right) . \left(A_x \hat{i} + A_y \hat{j} + A_z \hat{k} \right)$$

$$\vec{\nabla} . \vec{A} = \frac{\partial A_x}{\partial x} + \frac{\partial A_y}{\partial y} + \frac{\partial A_z}{\partial z} \qquad \qquad ...(1)$$

Divergence may be defined as the limiting values of the ratio of the flux of a vector across any closed surface around the point to the volume of the enclosure, when volume is contracted to zero.

i.e.
$$\vec{\nabla} . \vec{A} = \lim_{\Delta V \to 0} \frac{\oint_s \vec{A} . \overrightarrow{ds}}{\Delta V} \qquad \qquad ...(2)$$

Hence divergences of $\vec{A}$ is the net outward flux of the vector field $\vec{A}$ per unit volume as the volume shrinks to zero. Mathematically divergence is a measure of how much quantity comes out from some point.

The divergence of a vector field can also be viewed as the limit of the field's source strength per unit volume, which is positive at a point of source, negative at a point of sink and zero at where no source or sink.

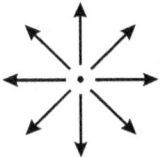

Source → positive
divergence

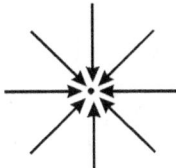

Sink → negative
divergence

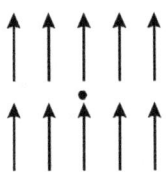

Zero divergence

If $\vec{\nabla}.\,\vec{v} = 0$ it implies physically that there is no inflow or outflow and the vector is called solenoidal vector. The equation $\vec{\nabla}.\,\vec{A} = 0$ is also known as continuity equation of an incompressible liquid.

Gauss Divergence Theorem

It states that the total outward flux of a vector field $\vec{A}$ through the closed surface S is equal to the volume integral of the divergence of $\vec{A}$ over the enclosed volume.

$$\oint \vec{A}.\overrightarrow{ds} = \int_V \vec{\nabla}.\vec{A}\ dV$$

i.e. $\oint$ flow out through the surface $= \int$ faucets within the volume.

It related the surface integral of a vector $\vec{A}$ with the volume integral of its divergence.

Divergence in cylindrical coordinate system

as
$$\text{div }\vec{A} = \vec{\nabla}.\vec{A}$$

$$= \left(\frac{\partial}{\partial\rho}\,\hat{\rho} + \frac{1}{\rho}\frac{\partial}{\partial\phi}\,\hat{\phi} + \frac{\partial}{\partial z}\,\hat{z}\right).\left(A_\rho\,\hat{\rho} + A_\phi\,\hat{\phi} + A_z\,\hat{z}\right)$$

$$= \frac{\partial A_\rho}{\partial\rho} + \frac{1}{\rho}\left(\frac{\partial\hat{\rho}}{\partial\phi}A_\rho + \frac{\partial A_\rho}{\partial\phi}\hat{\rho} + \frac{\partial\hat{\phi}}{\partial\phi}A_\phi + \hat{\phi}\frac{\partial A_\theta}{\partial\phi} + \hat{z}\frac{\partial A_z}{\partial\phi}\right) + \frac{\partial A_z}{\partial z} \quad ...(1)$$

But
$$\hat{\phi} = \frac{\partial\hat{\rho}}{\partial\phi}$$

and
$$-\hat{\rho} = \frac{\partial\hat{\phi}}{\partial\phi}$$

so
$$\vec{\nabla}.\vec{A} = \frac{\partial A_\rho}{\partial\rho} + \frac{\hat{\phi}}{\rho}\left(\hat{\phi}A_\rho + \hat{\rho}\frac{\partial A_\rho}{\partial\phi} - \hat{\rho}A_\phi + \hat{\phi}\frac{\partial A_\phi}{\partial\phi} + \hat{z}\frac{\partial A_z}{\partial\phi}\right) + \frac{\partial A_z}{\partial z}$$

$$= \frac{\partial A_\rho}{\partial\rho} + \frac{1}{\rho}A_\rho + \frac{1}{\rho}\frac{\partial A_\phi}{\partial\phi} + \frac{\partial A_z}{\partial z}$$

so
$$\vec{\nabla}.\vec{A} = \frac{1}{\rho}\frac{\partial}{\partial\rho}\left(\rho\,A_\rho\right) + \frac{1}{\rho}\frac{\partial A_\phi}{\partial\phi} + \frac{\partial A_z}{\partial z} \qquad\qquad ...(2)$$

Divergence in spherical coordinate system

$$\text{div } \vec{A} = \vec{\nabla}.\vec{A}$$

$$= \left(\frac{\partial}{\partial r}\hat{r} + \frac{1}{r}\frac{\partial}{\partial\theta}\hat{\theta} + \frac{1}{r\sin\theta}\frac{\partial}{\partial\phi}\hat{\phi}\right)\cdot\left(\hat{r}A_r + A_\theta\hat{\theta} + A_\phi\hat{\phi}\right)$$

as

$$\hat{\theta} = \frac{\partial\hat{r}}{\partial\theta}; \quad \hat{\phi}\sin\theta = \frac{\partial\hat{r}}{\partial\phi}; \quad -\hat{r} = \frac{\partial\hat{\theta}}{\partial\theta}$$

$$\hat{\phi}\cos\theta = \frac{\partial\hat{\theta}}{\partial\phi} \quad \text{and} \quad -\hat{r}\sin\theta - \hat{\theta}\cos\theta = \frac{\partial\hat{\phi}}{\partial\phi}$$

$$\boxed{\vec{\nabla}.\vec{A} = \frac{1}{r^2}\frac{\partial}{\partial r}\left(r^2 Ar\right) + \frac{1}{r\sin\theta}\frac{\partial}{\partial\theta}(\sin\theta\, A\theta) + \frac{1}{r\sin\theta}\frac{\partial A_\phi}{\partial\phi}}$$

Properties of divergences

(i)   It produces a scalar field.

(ii)   The divergence of a scalar field makes no sense.

(iii) $\vec{\nabla}.\left(\vec{A} + \vec{B}\right) = \vec{\nabla}.\vec{A} + \vec{\nabla}.\vec{B}$

(iv)   The vector field with zero divergence is called solenoid field.

Curl of a vector (rotation of a vector)

When del operator operates on a differentiable vector field with cross product, the obtained vector is called curl of the vector.

If $\vec{A}$ is a differentiable vector point function.

Then

$$\vec{\nabla} \times \vec{A} = \text{curl } \vec{A} = \begin{vmatrix} \hat{i} & \hat{j} & \hat{k} \\ \dfrac{\partial}{\partial x} & \dfrac{\partial}{\partial y} & \dfrac{\partial}{\partial z} \\ A_1 & A_2 & A_3 \end{vmatrix}$$

$$\vec{\nabla} \times \vec{A} = \left(\frac{\partial A_3}{\partial y} - \frac{\partial A_2}{\partial z}\right)\hat{i} - \left(\frac{\partial A_3}{\partial x} - \frac{\partial A_1}{\partial z}\right)\hat{j} + \left(\frac{\partial A_2}{\partial x} - \frac{\partial A_1}{\partial y}\right)\hat{k}$$

Physically it is the maximum circulation of $\vec{A}$ per unit area as the area tends to zero and whose direction is the normal direction of the area when the area is oriented to make the circulation maximum. i.e. it is the amount of maximum line integral at any point in a vector field per unit area around a closed curve.

So
$$\vec{\nabla} \times \vec{A} = \lim_{\Delta s \to 0} \frac{\oint \vec{A}.\,\overline{dl}}{\Delta s}$$

In hydrodynamics curl of a vector field at point indicates the amount of rotation of that vector about that point.

If $\vec{\nabla} \times \vec{A} = 0$ i.e. it is irrotational motion of incompressible fluid otherwise irrotational motion of compressible fluid.

If the field is not irrotational then it is called vortex field.

Divergence of a curl is always zero. $\vec{\nabla}.(\vec{\nabla} \times \vec{A}) = 0$.

Curl of a gradient is always zero $\vec{\nabla} \times \vec{\nabla} A = 0$

Properties

(i)   The curl of a vector is a vector

(ii) $\vec{\nabla} \times (\vec{A} \times \vec{B}) = \vec{A}(\vec{\nabla}.\vec{B}) - \vec{B}(\vec{\nabla}.\vec{A}) + (\vec{B}.\vec{\nabla}) - (\vec{A}.\vec{\nabla})\vec{A}$

(iii) $\vec{\nabla} \times (\vec{A} + \vec{B}) = \vec{\nabla} \times \vec{A} + \vec{\nabla} \times \vec{B}$

(iv) $\vec{\nabla} \times (\vec{A}\,\vec{B}) = \vec{A}\,\vec{\nabla} \times \vec{B} + \vec{\nabla}\,\vec{B} \times \vec{A}$

(v) $\vec{\nabla} \times (\vec{\nabla}\,\vec{B}) = 0$

(vi) $\vec{\nabla}.(\vec{\nabla} \times \vec{B}) = 0$

(vii) $\vec{\nabla} \times (\vec{\nabla} \times \vec{B}) = \vec{\nabla}(\vec{\nabla}.\vec{B}) - \vec{\nabla}^2\vec{B}$

(viii) $\vec{\nabla} \times \vec{B} = 0$ i.e. irrotational or conservative field.

(ix) $\text{curl}(\text{div } \vec{A}) = 0 \quad \vec{\nabla} \times (\vec{\nabla} \cdot \vec{A}) = 0$

Stoke's theorem

It states that the line integral of a vector around a closed curve is equal to the surface integral of the curl of that vector over that bounded by that closed curve.

i.e.
$$\oint_c \vec{A}.\,\overline{dl} = \oint_s (\vec{\nabla} \times \vec{A}).\,\overline{ds}$$

Stoke's theorem transforms the surface integral of the curl of a vector into the line integral of that vector over the boundary of the surface.

Curl in cylindrical coordinate system

$$\vec{\nabla} \times \vec{A} = \left(\hat{\rho}\frac{\partial}{\partial \rho} + \frac{1}{\rho}\frac{\partial}{\partial \phi}\hat{\phi} + \frac{\partial}{\partial z}\hat{z} \right) \times \left(A_\rho \hat{\rho} + A_\phi \hat{\phi} + A_z \hat{z} \right)$$

$$= \hat{\rho} \times \left[\frac{\partial A_\rho}{\partial \rho}\hat{\rho} + \frac{\partial A_\phi}{\partial \rho}\hat{\phi} + \frac{\partial A_z}{\partial \rho}\hat{z} \right]$$

$$+ \frac{\hat{\phi}}{\rho} \times \left[\frac{\partial \hat{\rho}}{\partial \phi}A_\rho + \frac{\partial A_\phi}{\partial \phi}\hat{\rho} + \frac{\partial \hat{\phi}}{\partial \phi}A_\phi + \frac{\partial A_\phi}{\partial \phi}\hat{\phi} + \frac{\partial A_z}{\partial \phi}\hat{z} \right]$$

$$+ \hat{z} \times \left[\frac{\partial A_\rho}{\partial z}\hat{\rho} + \frac{\partial A_\phi}{\partial z}\hat{\phi} + \frac{\partial A_z}{\partial z}\hat{z} \right]$$

$$= \left[\frac{\partial A_\phi}{\partial \rho}\hat{z} - \frac{\partial A_z}{\partial \rho}\hat{\phi} \right] + \frac{1}{\rho}\left[-\frac{\partial A_\rho}{\partial z}\hat{z} + A_\phi \hat{z} + \frac{\partial A_z}{\partial \phi}\hat{\rho} \right] + \left[\frac{\partial A_\rho}{\partial z}\hat{\phi} - \frac{\partial A_\phi}{\partial z}\hat{\rho} \right]$$

$$= \hat{\rho}\left[\frac{1}{\rho}\frac{\partial A_z}{\partial \phi} - \frac{\partial A_\phi}{\partial z} \right] + \hat{\phi}\left[\frac{\partial A_\rho}{\partial z} - \frac{\partial_z}{\partial \rho} \right] + \hat{z}\left[\frac{\partial A_\phi}{\partial \rho} - \frac{1}{\rho}\frac{\partial A_\phi}{\partial \phi} + \frac{A_\phi}{\rho} \right]$$

So

$$\vec{\nabla} \times \vec{A} = \hat{\rho}\left[\frac{1}{\rho}\frac{\partial A_z}{\partial \phi} - \frac{\partial A_\phi}{\partial z} \right] + \hat{\phi}\left[\frac{\partial A_\rho}{\partial z} - \frac{\partial A_z}{\partial \rho} \right] + \frac{\hat{z}}{\rho}\left[\frac{\partial}{\partial \rho}\left(\rho A_\phi \right) - \frac{\partial A_\rho}{\partial \phi} \right]$$

Curl in spherical coordinates

$$\vec{\nabla} \times \vec{A} = \left[\frac{\partial}{\partial r}\hat{r} + \frac{1}{r}\frac{r}{\partial \theta}\hat{\theta} + \frac{1}{r\sin\theta}\frac{\partial}{\partial \phi}\hat{\phi} \right] \times \left[A_r\hat{r} + A_\theta\hat{\theta} + A_\phi\hat{\phi} \right]$$

$$= \frac{\hat{r}}{r\sin\theta}\left[\frac{\partial}{\partial \theta}\left(\sin\theta\, A_\phi \right) - \frac{\partial A_\theta}{\partial \phi} \right] + \frac{\hat{\theta}}{r}\left[\frac{1}{\sin\theta}\frac{\partial A_r}{\partial \phi} - \frac{\partial}{\partial r}\left(rA_\phi \right) \right] + \frac{\hat{\phi}}{r}\left[\frac{\partial}{\partial r}\left(rA_\theta \right) - \frac{\partial A_r}{\partial \theta} \right]$$

The Laplacian

The laplacian of a scalar function A is the divergence of the gradient of A. The gradient of A is a vector and the divergence of vector is a scalar. Hence Laplacian gives a scalar.

Mathematically it is $\vec{\nabla}.\vec{\nabla}A = \nabla^2 A$ ..

In Cartesian coordinates–

$$\nabla^2 A = \vec{\nabla}.\vec{\nabla}A = \left(\frac{\partial}{\partial x}\hat{a}_x + \frac{\partial}{\partial y}\hat{a}_y + \frac{\partial}{\partial z}\hat{a}_z \right).\left(\frac{\partial A}{\partial x}\hat{a}_x + \frac{\partial A}{\partial y}\hat{a}_y + \frac{\partial A}{\partial z}\hat{a}_z \right)$$

$$\nabla^2 A = \frac{\partial^2 A}{\partial x} + \frac{\partial^2 A}{\partial y} + \frac{\partial^2 A}{\partial z} \qquad \qquad \dots(1)$$

In cylindrical coordinates

$$\nabla^2 A = \frac{1}{\rho} \cdot \frac{\partial}{\partial \rho}\left(\rho \frac{\partial A}{\partial \rho}\right) + \frac{1}{\rho^2}\frac{\partial^2 A}{\partial \phi^2} + \frac{\partial^2 A}{\partial z^2} \qquad \qquad \dots(2)$$

and in spherical coordinates

$$\nabla^2 A = \frac{1}{r^2} \cdot \frac{\partial}{\partial r}\left[r^2 \frac{\partial A}{\partial r}\right] + \frac{1}{r^2 \sin\theta}\frac{\partial}{\partial \theta}\left(\sin\theta \frac{\partial A}{\partial \theta}\right)$$

$$+ \frac{1}{r^2 \sin^2\theta}\frac{\partial^2 A}{\partial \phi^2} \qquad \qquad \dots(3)$$

if $\nabla^2 A = 0$ then the scalar field A is said to be harmonic.

$$\nabla^2 \vec{A} = \vec{\nabla}\left(\vec{\nabla}.\vec{A}\right) - \vec{\nabla}\times\vec{\nabla}\times\vec{A} \text{ if } \vec{A} \text{ is a vector.}$$

$$\nabla^2 \vec{A} = \nabla^2 A_x \hat{a}_x + \nabla^2 A_y \hat{a}_y + \nabla^2 A_z \hat{a}_z.$$

Line Integrals

It is the integral of the tangential component of any vector along the curve.

$$\int_L \vec{A}.\vec{dl} = \int_a^b |A| \cos\theta \, dl$$

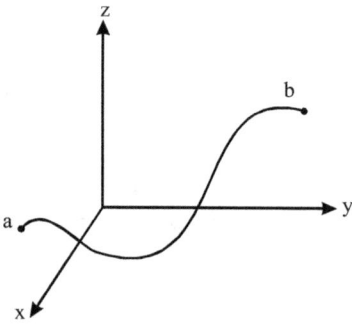

if it is a closed loop then

$$\oint_L \vec{A}.\vec{dl} = \int_a^b |A| \cos\theta \, dl$$

In Cartesian coordinate system

$$\int_a^b \vec{A}(x,y,z)\cdot \overrightarrow{dl} = \int_{x_a}^{x_b} A_x dx + \int_{y_a}^{y_b} A_y dy + \int_{z_a}^{z_b} A_z dz$$

in spherical system

$$\int_a^b \vec{A}(r,\theta,\phi)\cdot \overrightarrow{dl} = \int_{r_a}^{r_b} A_r dr + \int_{\theta_a}^{\theta_b} r A_\theta d\theta + \int_{\phi_a}^{\phi_b} r\sin\theta\, A_\phi d\phi$$

in cylindrical system

$$\int_a^b \vec{A}(\rho,\phi,z)\cdot \overrightarrow{dl} = \int_{\rho_a}^{\rho_b} A_\rho d\rho + \int_{\phi_a}^{\phi_b} r A_\phi d\phi + \int_{z_a}^{z_b} A_z dz$$

Surface integrals

It is the cross product or dot product of a vector $\vec{A}$ with a surfaces element $\overrightarrow{ds}$ for the entire closed surface S.

$$\oint \vec{A}\cdot\overrightarrow{ds} = \oiint_s \vec{A}\cdot\overrightarrow{ds} \text{ or } \oint_s \vec{A}\times\overrightarrow{ds}$$

physically it represents flux of vector $\vec{A}$ over the entire surface s.

e.g. the surfaces integral of an electric field $\vec{E}$ over the surface s represents the flux of $\vec{E}$.

i.e. flux $\phi = \oint_s \vec{E}\cdot\overrightarrow{ds}$

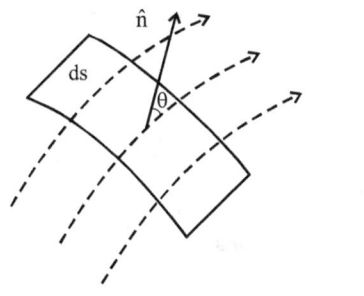
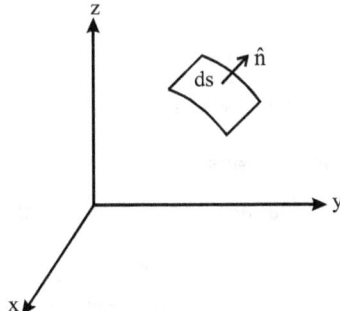

Volume Integral

It is the integral of the scalar function over the entire volume.

$$\int_V \rho \, dV \quad \text{Where } \rho \text{ is a scalar function and } dV \text{ is infinitesimal volume}$$

element.

Physical significance depends on the nature of physical quantity for ρ. for example if ρ is density the volume integral gives the total mass and if ρ represents the charge density then volume integral gives total charge bounded by the surfaces.

Some important results involving del operator

(1) $\quad \vec{\nabla}(A+B) = \vec{\nabla}A + \vec{\nabla}B$

(2) $\quad \vec{\nabla} \cdot (\vec{A}+\vec{B}) = \vec{\nabla} \cdot \vec{A} + \vec{\nabla} \cdot \vec{B}$

(3) $\quad \vec{\nabla} \times (\vec{A}+\vec{B}) = \vec{\nabla} \times \vec{A} + \vec{\nabla} \times \vec{B}$

(4) $\quad \vec{\nabla} \cdot (\phi \vec{A}) = \vec{\nabla}\phi \cdot \vec{A} + \phi(\vec{\nabla} \cdot \vec{A})$

(5) $\quad \vec{\nabla} \times (\phi \vec{A}) = \vec{\nabla}\phi \times \vec{A} + \phi(\vec{\nabla} \times \vec{A})$

(6) $\quad \vec{\nabla} \cdot (\vec{A} \times \vec{B}) = \vec{B} \cdot (\vec{\nabla} \times \vec{A}) - \vec{A} \cdot (\vec{\nabla} \times \vec{B})$

(7) $\quad \vec{\nabla} \times (\vec{A} \times \vec{B}) = (\vec{B} \cdot \vec{\nabla})\vec{A} - (\vec{\nabla} \cdot \vec{A})\vec{B} - (\vec{A} \cdot \vec{\nabla})\vec{B} + (\vec{\nabla} \cdot \vec{B})\vec{A}$

(8) $\quad \vec{\nabla}(\vec{A} \cdot \vec{B}) = (\vec{B} \cdot \vec{\nabla})\vec{A} + (\vec{A} \cdot \vec{\nabla})\vec{B} + \vec{B} \times (\vec{\nabla} \times \vec{A}) + \vec{A} \times (\vec{\nabla} \times \vec{B})$

(9) $\quad \vec{\nabla}(\vec{\nabla}\phi) = \nabla^2\phi = \dfrac{\partial^2\phi}{\partial x^2} + \dfrac{\partial^2\phi}{\partial y^2} + \dfrac{\partial^2\phi}{\partial z^2}$

(10) $\quad \vec{\nabla} \times (\vec{\nabla}\phi) = 0$

(11) $\quad \vec{\nabla} \cdot (\vec{\nabla} \times \vec{A}) = 0$

(12) $\quad \vec{\nabla} \times (\vec{\nabla} \times \vec{A}) = \vec{\nabla}(\vec{\nabla} \cdot \vec{A}) - \nabla^2\vec{A}$

Types of vector fields

(i) Solenoidal and Irrotational field (Lamellar)

if curl $\vec{R} = 0 \Rightarrow \vec{R} = \text{grad } \mu$ where μ is the scalar potential.

$\therefore \qquad \qquad \text{div grad } \mu = \nabla^2 \mu = 0 \ (\text{given div } \vec{R} = 0)$

This equation is known as Lapalce's equation and such fields are called Laplacians. e.g. Electrostatic field in free space, gravitational field in free space, thermal fields in equilibrium, magnetostatic fields in current free region, static current field within a linear homogenous isotropic conductor.

(ii) Irrotational but not solenoidal field

Here curl $\vec{R} = 0$ but div $\vec{R} \neq 0$

again with $\vec{R}$ = grad x, x being the scalar potential but div grad x = $\nabla^2 x \neq 0$

This is called the Poisson's equation and such fields are known as poissonian. e.g. electrostatic fields in a charged medium, electrons inside a thermionic tube, gravitational force inside a mass.

(iii) Solenoidal but not irrotational field

here div $\vec{R} = 0,$ but curl $\vec{R} \neq 0$

since curl $\vec{R} \neq 0$ $\vec{R}$ = curl ϕ where ϕ is the vector potential

∴   div R = div curl $\phi = 0$ (A vector identity)

and curl $\vec{R}$ = curl $\phi \neq 0$

$$= \text{grad div } \phi - \nabla^2 \phi \neq 0$$

if now   div $\phi = 0$ then $\nabla^2 \phi \neq 0$

This is similar to Poisson's equation but it is terms of a vector potential. e.g. magnetic field within a conductor carrying a steady current, Rotational motion of an incompressible fluid, time varying electromagnetic field in charge free and current free region.

Neither irrotational nor solenoidal field

for this

$$\text{curl } \vec{R} \neq 0 \quad \text{and div } \vec{R} \neq 0$$
$$\vec{R} = \text{grad x + curl } \phi$$
$$x = \text{scalar potential}$$
$$\phi = \text{vector potential}$$

∴   div $\vec{R}$ = div grad x + div curl $\phi \neq 0$

But   div curl $\phi = 0$

so   div grad x $\neq 0$ This is Poisson's equation

and   curl $\vec{R}$ = curl grad x + curl curl ϕ

But curl grad x = 0

∴   curl curl $\phi \neq 0$

It can be reduced to $\nabla^2\phi \neq 0$ by assuming div $\phi = 0$. This is the most general type of field.

It can be decomposed into two fields - one is lamellar with x and another solenoidal with ϕ. this is known as 'Helmholz theorem' e.g. Rotational motion of a compressible liquid. Helmholz theorem states that a vector function is determinal uniquely if the values of its curl of divergence are known at all points.

SOLVED EXAMPLES

Q.1 If vector $\vec{A} = 2\hat{i} - \hat{j} + 2\hat{k}$, $\vec{B} = 2\hat{i} - \hat{j}$, $\vec{C} = 2\hat{i} - 3\hat{j} + \hat{k}$ **then find**

(a) $\vec{A} + \vec{B}$ (b) $\vec{A} - \vec{B}$ (c) $\vec{A} \cdot (\vec{B} \times \vec{C})$ (d) $\vec{B} \cdot (\vec{C} \times \vec{A})$

(e) $\vec{A} \times (\vec{B} \times \vec{C})$ (f) a unit vector perpendicular to both $\vec{B}$ and $\vec{C}$

(g)   Component of $\vec{A}$ along $\vec{B}$.

Solution:

(a)
$$\vec{A} + \vec{B} = \left(2\hat{i} - \hat{j} + 2\hat{k}\right) + \left(2\hat{i} - \hat{j}\right)$$
$$= 4\hat{i} - 2\hat{j} + 2\hat{k}$$

(b)
$$\vec{A} - \vec{B} = \left(2\hat{i} - \hat{j} + 2\hat{b}\right) + \left(2\hat{i} - \hat{j}\right)$$
$$= 2\hat{k}$$

(c)
$$\vec{A} \cdot (\vec{B} \times \vec{C}) = \begin{vmatrix} 2 & -1 & 2 \\ 2 & -1 & 0 \\ 2 & -3 & 1 \end{vmatrix}$$
$$= -2 + 2 - 8 = -8$$

(d)
$$\vec{B} \cdot (\vec{C} \times \vec{A}) = \begin{vmatrix} 2 & -1 & 0 \\ 2 & -3 & 1 \\ 2 & -1 & 2 \end{vmatrix}$$
$$= -10 + 2$$
$$= -8$$

(e)
$$\vec{A} \times (\vec{B} \times \vec{C}) = \vec{B}(\vec{A} \cdot \vec{C}) - \vec{C}(\vec{A} \cdot \vec{B})$$
$$= \left(2\hat{i} - \hat{j}\right)\left[\left(2\hat{i} - \hat{j} + 2\hat{k}\right) \cdot \left(2\hat{i} - 3\hat{j} + \hat{k}\right)\right]$$

$$-\left(2\hat{i}-3\hat{j}+\hat{k}\right)\left[\left(2\hat{i}-\hat{j}+2\hat{k}\right).\left(2\hat{i}-\hat{j}\right)\right]$$

$$= 12\hat{i}-3\hat{k}$$

(f)

$$\hat{n} = \frac{\vec{B}\times\vec{C}}{\left|\vec{B}\times\vec{C}\right|} = \frac{\left(2\hat{i}-\hat{j}\right)\times\left(2\hat{i}-3\hat{j}+\hat{k}\right)}{\sqrt{(-1)^2+(-2)^2+(-4)^2}}$$

$$= \frac{-\hat{i}-2\hat{j}-4\hat{b}}{\sqrt{21}}$$

(g)

$$A_B = |A|\cos\theta\,\hat{B}$$

$$= \left(\vec{A}.\hat{B}\right)\hat{B} = \frac{\left(\vec{A}.\vec{B}\right)\vec{B}}{|B|^2} \quad\text{as}\quad \hat{B} = \frac{\vec{B}}{|B|}$$

$$= \frac{\left[\left(2\hat{i}-\hat{j}+2k\right).\left(2\hat{i}-\hat{j}\right)\right]\left(2\hat{i}-\hat{j}\right)}{\left(2^2+(-1)^2\right)}$$

$$= \frac{10\hat{i}-5\hat{j}}{5}$$

$$= 2\hat{i}-\hat{j}$$

Q.2: Find the angle between $\vec{A} = 2\hat{i}-\hat{j}+\hat{k}$ and $\vec{B} = \hat{i}+\hat{j}+3\hat{k}$.

Solution :

as

$$\vec{A}.\vec{B} = \left|\vec{A}\right|\left|\vec{B}\right|\cos\theta$$

$$\cos\theta = \frac{\vec{A}.\vec{B}}{\left|\vec{A}\right|\left|\vec{B}\right|} = \frac{\left(2\hat{i}-\hat{j}+\hat{k}\right).\left(\hat{i}+\hat{j}+3\hat{k}\right)}{\sqrt{2^2+(-1)^2+1^2}\,\sqrt{1^2+1^2+3^2}}$$

$$\theta = \cos^{-1}\frac{4}{\sqrt{6}\,\sqrt{11}}$$

$$= 60.5° \text{ approx.}$$

Q.3: If $\vec{A} = 2\hat{i}-3\hat{j}+7\hat{k}$ and $\vec{B}=-2\hat{i}+\hat{j}+\hat{k}$, then show that $\vec{A}$ and $\vec{B}$ are perpendicular to each other.

Solution :

$$\vec{A}.\vec{B} = \left(2\hat{i}-3\hat{j}+7\hat{k}\right).\left(-2\hat{i}+\hat{j}+\hat{k}\right)$$

$$= -4 - 3 + 7 = 0$$

so $$\vec{A}.\vec{B} = \left|\vec{A}\right|\left|\vec{B}\right|\cos\theta = 0$$

$$\Rightarrow \qquad\qquad \cos\theta = 0$$

$$\Rightarrow \qquad\qquad \theta = 90°$$

so $\vec{A}$ is perpendicular to $\vec{B}$.

Q.4: Find the unit vector perpendicular to both

$\vec{A} = 2\hat{i} + 3\hat{j} - 5\hat{k}$ **and** $\vec{B} = 2\hat{i} + 3\hat{j} - \hat{k}$. **Also find the angle between them.**

Solution:

As $$\left(\vec{A} \times \vec{B}\right) = \begin{vmatrix} \hat{i} & \hat{j} & \hat{k} \\ 2 & 3 & -5 \\ 2 & 3 & -1 \end{vmatrix}$$

$$= 12\hat{i} - 8\hat{j}$$

unit vector perpendicular to both $\vec{A}$ and $\vec{B}$

$$\hat{n} = \frac{\vec{A} \times \vec{B}}{\left|\vec{A} \times \vec{B}\right|}$$

$$= \frac{12\hat{i} - 8\hat{j}}{\sqrt{12^2 + (-8)^2}} = \frac{12\hat{i} - 8\hat{j}}{\sqrt{208}}$$

Again $$\sin\theta = \frac{\left|\vec{A} \times \vec{B}\right|}{\left|\vec{A}\right|\left|\vec{B}\right|}$$

$$= \frac{\sqrt{208}}{\sqrt{4+9+25}\sqrt{4+9+1}} = \frac{\sqrt{208}}{\sqrt{38}\sqrt{14}}$$

$$\theta = \sin^{-1}\left(\frac{\sqrt{208}}{\sqrt{38}\sqrt{14}}\right)$$

$$= 38.7° \text{ Approx.}$$

Q.5: A particle is acted upon by two constant forces $\vec{F}_1 = \hat{i} + 4\hat{j} - 3\hat{k}$ **and** $\vec{F}_2 = 3\hat{i} + \hat{j} - \hat{k}$ **due to which particle is displaced from** $\hat{i} + 2\hat{j} + 3\hat{k}$ **to** $4\hat{i} + 5\hat{j} + \hat{k}$. **Calculate the total work done.**

Solution:

Displacement of the particle

$$\vec{r} = 4\hat{i} + 5\hat{j} + \hat{k} - \left(\hat{i} + 2\hat{j} + 3\hat{k}\right)$$

$$= 3\hat{i} + 3\hat{j} - 2\hat{k}$$

Hence total work done

$$= \text{Total force . displacement}$$

$$= \left(\vec{F}_1 + \vec{F}_2\right).\vec{r}$$

$$= \left[\left(\hat{i} + 4\hat{j} - 3\hat{k}\right) + \left(3\hat{i} + \hat{j} - \hat{k}\right)\right].\left(3\hat{i} + 3\hat{j} - 2\hat{k}\right)$$

$$= \left(4\hat{i} + 5\hat{j} - 4\hat{k}\right).\left(3\hat{i} + 3\hat{j} - 2\hat{k}\right) = 12 + 15 + 8$$

$$= 35 \text{ units.}$$

Q.6: A rigid body is rotating with angular velocity of 5 rad/s about an axis parallel to $3\hat{j} - \hat{k}$ and passing through the point $\hat{i} + \hat{j} - 3\hat{k}$. Find the velocity vector of the particle, when it is at the point $2\hat{i} - 4\hat{j} + \hat{k}$.

Solution : Suppose $\vec{r}$ is the position vector

then

$$\vec{r} = \left(2\hat{i} - 4\hat{j} + \hat{k}\right) - \left(\hat{i} + \hat{j} - 3\hat{k}\right)$$

$$= \hat{i} - 5\hat{j} + 4\hat{k}$$

angular velocity

$$\vec{\omega} = 5 \times \frac{\left(3\hat{j} - \hat{k}\right)}{\left|3\hat{j} - \hat{k}\right|} \frac{5}{\sqrt{10}} \left(3\hat{j} - \hat{k}\right)$$

linear velocity

$$\vec{v} = \vec{\omega} \times \vec{r} = \left(\frac{5}{\sqrt{10}} 3\hat{j} - \hat{k}\right) \times \left(\hat{i} - 5\hat{j} + 4\hat{k}\right)$$

$$= \frac{5}{\sqrt{10}} \begin{vmatrix} \hat{i} & \hat{j} & \hat{k} \\ 0 & 3 & -1 \\ 1 & -5 & 4 \end{vmatrix}$$

$$= \frac{5}{\sqrt{10}} \left(7\hat{i} - \hat{j} - 3\hat{k}\right) \text{ units.}$$

Q.7 : Calculate the torque of a force $-2\hat{i}+2\hat{j}+5\hat{k}$ about the point $8\hat{j}$ acting through the point $6\hat{i}+4\hat{j}+2\hat{k}$.

Solution : Here

$$\vec{r} = 8\hat{j}-\left(6\hat{i}+4\hat{j}+2\hat{k}\right)$$

$$= -6\hat{i}+4\hat{j}-2\hat{k}$$

$$\text{torque } \tau = \vec{r}\times\vec{F} = \left(-6\hat{i}+4\hat{j}-2\hat{k}\right)\times\left(-2\hat{i}+2\hat{j}+5\hat{k}\right)$$

$$= \begin{vmatrix} \hat{i} & \hat{j} & \hat{k} \\ -6 & 4 & -2 \\ -2 & 2 & 5 \end{vmatrix}$$

$$= 24\hat{i}+34\hat{j}-4\hat{k}$$

Q.8: A force vector $10\hat{i}+25\hat{j}+35\hat{k}$ passes through a point (2, 5, 7). Prove that force is also passing through the origin.

Solution: The position vector

$$\vec{r} = 2\hat{i}+5\hat{j}+7\hat{k}$$

and moment of the form about this point i.e. torque

$$\tau = \vec{r}\times\vec{F}$$

$$= \left(2\hat{i}+5\hat{j}+7\hat{k}\right)\times\left(10\hat{i}+25\hat{j}+35\hat{k}\right)$$

$$= \begin{vmatrix} \hat{i} & \hat{j} & \hat{k} \\ 2 & 5 & 7 \\ 10 & 25 & 35 \end{vmatrix}$$

$$= 0$$

As the moment is zero, which shows that forces is passing through the origin.

Q.9: A force $4\hat{i}-3\hat{j}+2\hat{k}$ passes through the point (–9, 2, 1). Find the component of moment of the force about the axis of reference.

Sol.: Here

$$\vec{r} = -9\hat{i}+2\hat{j}+\hat{k}$$

so moment of force i.e. torque

$$\vec{\tau} = \vec{r} \times \vec{F} = \left(-9\hat{i} + 2\hat{j} + \hat{k}\right) \times \left(4\hat{i} - 3\hat{j} + 2\hat{k}\right)$$

$$= \begin{vmatrix} \hat{i} & \hat{j} & \hat{k} \\ -9 & 2 & 1 \\ 4 & -3 & 2 \end{vmatrix}$$

$$= 7\hat{i} + 22\hat{j} + 19\hat{k}$$

Hence components of moment of force are 7 unit, 22 units and 19 units in x, y and z direction respectively.

Q.10: A proton is moving with velocity 10^8 cm/s along z-axis through an electric field of intensity 3×10^4 volt/cm along x-axis and magnetic field of intensity 2000 gauss along y-axis. Calculate the magnitude and direction of total force.

Solution: Intensity of electric field

$$\vec{E} = 3 \times 10^4 \, \hat{i} \text{ volt / cm} = 100 \, \hat{i} \text{ esu/cm}$$

$$\text{Proton charge} = 1.6 \times 10^{-19} \, C = 4.8 \times 10^{-10} \text{ esu}$$

$$\text{Magnetic field } \vec{B} = 2000 \, \hat{j} \text{ gauss}$$

$$\text{velocity } \vec{v} = 10^8 \, \hat{k} \text{ cm / s}$$

so total force acting on the proton

$$\vec{F} = q\left(\vec{E} + \frac{\vec{v} \times \vec{B}}{C}\right)$$

$$= \left(4.8 \times 10^{-10}\right)\left[100\hat{i} + \frac{1}{3 \times 10^{10}}\left\{10^8 \, \hat{k} \times 2000\hat{j}\right\}\right]$$

$$= 4.47 \times 10^{-8} \hat{i} \text{ dyne}$$

Hence total force acting on the proton has magnitude +4.47 × 10⁻⁸ dyne along the +ve x-direction.

Q.11: Find the value of the constant p so that

$$\vec{A} = 2\hat{i} + \hat{j} - 3\hat{k}, \ \vec{B} = 2\hat{i} - 3\hat{j} - \hat{k} \text{ and } \vec{C} = 3\hat{i} - p\hat{j} + \hat{k} \text{ are coplanar.}$$

Solution: We know that three vectors are said to be coplanar if $\vec{A}.\left(\vec{B} \times \vec{C}\right) = 0$

$$\therefore \qquad \begin{vmatrix} 2 & 1 & -3 \\ 2 & -3 & -1 \\ 3 & -p & 1 \end{vmatrix} = 0$$

$$\Rightarrow \qquad 4p - 38 = 0$$

or $\qquad 4p = 38$

$$p = \frac{38}{4} = 9.5$$

Q.12: Evaluate $\vec{A} \times (\vec{B} \times \vec{C})$ where

$$\vec{A} = 2\hat{i} + \hat{j}, \quad \vec{B} = -\hat{i} + \hat{j} + \hat{k} \text{ and } \quad \vec{C} = 5\hat{i} - 3\hat{j} + \hat{k}.$$

Solution:

$$\vec{A} \times (\vec{B} \times \vec{C}) = \vec{B}(\vec{A}.\vec{C}) - \vec{C}(\vec{A}.\vec{B})$$

$$= (-\hat{i} + \hat{j} + \hat{k})\left\{(2\hat{i} + \hat{j} + 0\hat{k}).(5\hat{i} - 3\hat{j} + \hat{k})\right\}$$

$$- (5\hat{i} - 3\hat{j} + \hat{k})\left\{(2\hat{i} + \hat{j} + 0\hat{k}).(-\hat{i} + \hat{j} + \hat{k})\right\}$$

$$= 7(-\hat{i} + \hat{j} + \hat{k}) - (1)(5\hat{i} - 3\hat{j} + \hat{k})$$

$$= -2\hat{i} + 4\hat{j} + 8\hat{k}$$

Q.13: If $\vec{r}$ si the position vector of any point (x, y, z) and $\vec{A}$ is a constant vector then show that

(i) $\quad (\vec{r}.\vec{A}).\vec{A} = 0$ is the equation of a constant plane.

(ii) $\quad (\vec{r} - \vec{A}).\vec{r}$ is the equation of a sphere.

Also show that result of (i) is of the form $Ax + By + Cz + D = 0$ where $D = -(A^2 + B^2 + C^2)$ and that of (ii) is of the from $x^2 + y^2 + z^2 = r^2$. [RU 2005]

Solution: (i) Suppose $\vec{A} = (A, B, C)$ and $\vec{r} = (x, y, z)$

$$(\vec{r} - \vec{A}).\vec{A} = (x - A)A + (y - B)B + (z - C)C$$

$$= xA - A^2 + yB - B^2 + zC - C^2$$

$$= xA + yB + zC - \left(A^2 + B^2 + C^2\right)$$

$$= A_x + B_y + C_z + D$$

where
$$D = -\left(A^2 + B^2 + C^2\right)$$

so
$$\left(\vec{r} - \vec{A}\right).\vec{A} = 0 \rightarrow A_x + B_y + C_z + D = 0$$

which is an equation of a plane.

(ii)
$$\left(\vec{r} - \vec{A}\right).\vec{A} = (x - A)x + (y - B)y + (z - C)z$$

if
$$\left(\vec{r} - \vec{A}\right).\vec{A} = 0 \text{ then}$$

$$x^2 + y^2 + z^2 - A_x - B_y - C_z = 0$$

Which is the equation of sphere whose surface touches the origin.

Q.14: A particle moves on the curve $x = 2t^2$, $y = t^2 - 4t$, $z = 3t - 5$ where t is the time. Find the components of velocity and acceleration at time t = 1 in the direction $\hat{i} - 3\hat{j} + 2\hat{k}$.

Solution: Position vector

$$\vec{r} = 2t^2\hat{i} + \left(t^2 - 4t\right)\hat{j} + (3t - 5)\hat{k}$$

so velocity vector

$$\vec{v} = \frac{d\vec{r}}{dt} = \frac{d}{dt}\left(2t^2\right)\hat{i} + \frac{d}{dt}\left(t^2 - 4t\right)\hat{j} + \frac{d}{dt}(3t - 5)\hat{k}$$

$$= 4t\,\hat{i} + (2t - 4)\hat{j} + 3\hat{k}$$

acceleration
$$\vec{a} = \frac{d\vec{v}}{dt}$$

$$= \frac{d}{dt}(4t)\hat{i} + \frac{d}{dt}(2t - 4)\hat{j} + \frac{d}{dt}(3)\hat{k}$$

$$= 4\hat{i} + 2\hat{j} + 0$$

at
t = 1, velocity $\vec{v} = 4\hat{i} - 2\hat{j} + 3\hat{k}$

acceleration $\vec{a} = 4\hat{i} + 2\hat{j}$

and the component of $\vec{v}$ along $\hat{i} - 3\hat{j} + 2\hat{k}$

is $\dfrac{\left(4\hat{i} - 2\hat{j} + 3\hat{k}\right).\left(\hat{i} - 3\hat{j} + 2\hat{k}\right)}{\sqrt{1^2 + (-3)^2 + 2^2}}$

$$= \dfrac{16}{\sqrt{14}} = \dfrac{8\sqrt{14}}{7}$$

and component of $\vec{a}$ along $i - 3\hat{j} + 2\hat{k}$ is

$$= \dfrac{\left(4\hat{i} + 2\hat{j}\right).\left(\hat{i} - 3\hat{j} + 2\hat{k}\right)}{\sqrt{1^2 + (-3)^2 + 2^2}} = \dfrac{-2}{\sqrt{14}}$$

$$= \dfrac{-\sqrt{14}}{7}$$

Q.15: **Calculate the unit vector, which is normal to the surface**

$$\phi = x^2 y + xy^2 + 3xyz \text{ at the point } (1,\ 1,\ -1).$$

Solution : Here

$$\vec{\nabla}\phi = \left(\dfrac{\partial}{\partial x}\hat{i} + \dfrac{\partial}{\partial y}\hat{j} + \dfrac{\partial}{\partial z}\hat{k}\right)\left(x^2 y + xy^2 + 3xyz\right)$$

$$= \dfrac{\partial}{\partial x}\left(x^2 y + xy^2 + 3xyz\right)\hat{i} + \dfrac{\partial}{\partial y}\left(x^2 y + xy^2 + 3xyz\right)\hat{j}$$

$$+ \dfrac{\partial}{\partial z}\left(x^2 y + xy^2 + 3xyz\right)\hat{k}$$

$$= \left(2xy + y^2 + 3yz\right)\hat{i} + \left(x^2 + 2xy + 3xz\right)\hat{j} + \left(3xy\right)\hat{k}$$

At $(1,\ 1,\ -1)$,

$$\vec{\nabla}\phi = \left(2 + 1 - 3\right)\hat{i} + \left(1 + 2 - 3\right)\hat{j} + 3\hat{k} = 3\hat{k}$$

so the unit vector normal to the surface ϕ at $(1,\ 1,\ -1)$ is

$$\dfrac{3\hat{k}}{\sqrt{3^2}} = \dfrac{3\hat{k}}{3}$$

$$= \hat{k}$$

Q.16: Find the direction derivative of $\phi(x,y,z) = x^2y + xy^2$ at the point (2, –1, –4) along the direction of the vector (1, 2, –1).

Solution: as

$$\phi = x^2y + xy^2$$

$$\vec{\nabla}\phi = \left(\frac{\partial}{\partial x}\hat{i} + \frac{\partial}{\partial y}\hat{j} + \frac{\partial}{\partial z}\hat{k}\right)\phi(x,y,z)$$

$$= \frac{\partial}{\partial x}\left(x^2y + xy^2\right)\hat{i} + \frac{\partial}{\partial y}\left(x^2y + xy^2\right)\hat{j} + \frac{\partial}{\partial z}\left(x^2y + xy^2\right)\hat{k}$$

$$= \left(2xy + y^2\right)\hat{i} + \left(x^2 + 2xy\right)\hat{j}$$

$$\left(\vec{\nabla}\phi\right)_{(2,-1,-4)} = -3\hat{i}$$

Position vector $\hat{r} = \hat{i} + 2\hat{j} - \hat{k}$

and unit vector along this position vector

$$\hat{n} = \frac{\hat{i} + 2\hat{j} - \hat{k}}{\sqrt{1+4+1}} = \frac{\hat{i} + 2\hat{j} - \hat{k}}{\sqrt{6}}$$

and direction derivative $\vec{\nabla}\phi.\hat{n} = \left(-3\hat{i}\right).\dfrac{\left(\hat{i} + 2\hat{j} - \hat{k}\right)}{\sqrt{6}}$

$$= \frac{-3}{\sqrt{6}} = \frac{-\sqrt{6}}{2}$$

Q.17: Find the equation of the tangent plane and normal line to the surface $2x^2 + y^2 + 2z = 3$ at the point (2, 1, –3).

Solution :

Here $\qquad \phi(x,y,z) = 2x^2 + y^2 + 2z - 3$

$\therefore \qquad \dfrac{\partial\phi}{\partial x} = \dfrac{\partial}{\partial x}\left(2x^2 + y^2 + 2z\right) = 4x$

$\qquad \dfrac{\partial\phi}{\partial y} = \dfrac{\partial}{\partial y}\left(2x^2 + y^2 + 2z\right) = 2y$

$\qquad \dfrac{\partial\phi}{\partial z} = \dfrac{\partial}{\partial z}\left(2x^2 + y^2 + 2z\right) = 2$

so the components $\dfrac{\partial \phi}{\partial x}, \dfrac{\partial \phi}{\partial y}$ and $\dfrac{\partial \phi}{\partial z}$ at the point (2, 1, –3) will be

$$\dfrac{\partial \phi}{\partial x} = 4 \times 2 = 8, \quad \dfrac{\partial \phi}{\partial y} = 2 \times 1 = 2, \quad \dfrac{\partial \phi}{\partial z} = 2$$

Hence the equation of the tangent plane to the surface at the point (2, 1, –3) is

$$(X - 2)8 + (Y - 1)2 + (Z + 3)2 = 0$$

or $\qquad\qquad\qquad 4X + y + Z = 6$

so the equation of normal to the surface at (2, 1, –3) is

$$\dfrac{X - 2}{8} = \dfrac{Y - 1}{2} = \dfrac{Z + 3}{2}$$

or $\qquad\qquad\qquad \dfrac{X - 2}{4} = Y - 1 = Z + 3$

Q.18: Find the angle between the surfaces $x^2 + y^2 + z^2 = 9$ and $x^2 + y^2 - z = 3$ at the (1, 2, 2)

Solution:

Suppose $\qquad\qquad \phi_1 = x^2 + y^2 + z^2$ and $\phi_2 = x^2 + y^2 - z$

so $\qquad\qquad \vec{\nabla}\phi_1 = 2x\hat{i} + 2y\hat{j} + 2z\hat{k}$

and $\qquad\qquad \nabla\phi_2 = 2x\hat{i} + 2y\hat{j} - \hat{k}$

and $\qquad\qquad \left(\vec{\nabla}\phi_1\right)_{1,2,2} = 2\hat{i} + 4\hat{j} + 4\hat{k}$

$$\left(\vec{\nabla}\phi_2\right)_{1,2,2} = 2\hat{i} + 4\hat{j} - \hat{k}$$

since $\vec{\nabla}\phi_1$ and $\vec{\nabla}\phi_2$ are normal to ϕ_1 and ϕ_2

then $\qquad\qquad \vec{\nabla}\phi_1 . \vec{\nabla}\phi_2 = \left|\vec{\nabla}\phi_1\right|\left|\vec{\nabla}\phi_2\right| \cos\theta$ where θ is the angle between the surfaces ϕ_1 and ϕ_2.

so $\qquad\qquad \theta = \cos^{-1}\left[\dfrac{\vec{\nabla}\phi_1 . \vec{\nabla}\phi_2}{\left|\vec{\nabla}\phi_1\right|\left|\vec{\nabla}\phi_2\right|}\right]$

$$= \cos^{-1}\dfrac{4 + 16 - 4}{\sqrt{36 \times 21}} = \cos^{-1}\left(\dfrac{16}{6\sqrt{21}}\right)$$

$$\theta = 54.41° \text{ approx.}$$

Q.19 (i) Provle that $\vec{P} = \cos\theta_1\ \hat{i} + \sin\theta_1\hat{j}$ and $\cos\theta_2\hat{i} + \sin\theta_2\hat{j}$ are unit vectors in the xy-plane respectively making θ_1 and θ_2 with the x-axis.

(ii) By means of dot product, obtain the formula for $\cos(\theta_2 - \theta_1)$. by similarly formulating P and Q, obtain the formula for $\cos(\theta_2 + \theta_1)$.

(iii) If θ is the angle between P and Q find $\dfrac{1}{2}|P - Q|$ in terms of θ.

Solution:

(i)　　Given
$$\vec{P} = \cos\theta_1\hat{i} + \sin\theta_1\hat{j}$$
$$\vec{Q} = \cos\theta_2\hat{i} + \sin\theta_2\hat{j}$$

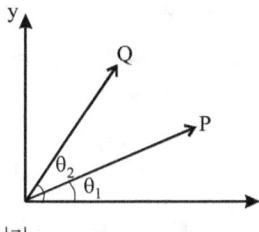

$$|\vec{P}| = \cos^2\theta_1 + \sin^2\theta_1 = 1$$
$$|\vec{Q}| = \cos^2\theta_2 + \sin^2\theta_2 = 1$$

hence $\vec{P}$ and $\vec{Q}$ are unit vectors.

(ii)
$$\vec{P}.\vec{Q} = |\vec{P}||\vec{Q}| \cos(\theta_2 - \theta_1)$$
$$= 1.1 \cos(\theta_2 - \theta_1) \qquad\qquad ...(1)$$

But
$$\vec{P}.\vec{Q} = (\cos\theta_1\hat{i} + \sin\theta_1\hat{j}).(\cos\theta_2\hat{i} + \sin\theta_2\hat{j})$$
$$= \cos\theta_1\cos\theta_2 + \sin\theta_1\sin\theta_2 \qquad\qquad ...(2)$$

so
$$\cos(\theta_2 - \theta_1) = \cos\theta_1\cos\theta_2 + \sin\theta_1\sin\theta_2$$

let
$$\vec{P}_1 = \vec{P} = \cos\theta_1\hat{i} + \sin\theta_1\hat{j}$$

and
$$\vec{Q}_1 = \cos\theta_2\hat{i} - \sin\theta_2\hat{j}$$

then $\vec{P}_1 . \vec{Q}_1 = 1.1 \cos (\theta_1 - \theta_2)$

$$= \cos \theta_1 \cos \theta_2 - \sin \theta_1 \sin \theta_2$$

(iii) $\vec{P}_1$ and $\vec{Q}_1$ are unit vectors.

so $\dfrac{1}{2} (\vec{P} - \vec{Q}) = \dfrac{1}{2} \left| \rho^2 + Q^2 - 2PQ \cos \theta \right|$

$$= \dfrac{1}{2} \left| 1 + 1 - 2 \cos \theta \right|$$

$$= 1 - \cos \theta$$

$$= 2 \sin^2 \dfrac{\theta}{2}$$

Q.20: A vector field is given as $\vec{W} = 4x^2y\,\hat{i} - (7x + 2z)\hat{j} + (4xy + 2z^2)\hat{k}$

(i)   **What is the magnitude of the field at point (2. –3, 4).**

(ii)   **At what point on z-axis is the magnitude of W equal to unity?** **[RU 2002]**

Solution: (i)

$$\vec{W} = 4x^2y\,\hat{i} - (7x + 2z)\hat{j} + (4xy + 2z^2)\hat{k}$$

at P(2, –3, 4), $\vec{W} = 4(2^2)(-3)\hat{i} - (7 \times 2 + 4 \times 2)\hat{j} + (4 \times 2 \times (-3) + 2 \times 4^2)\hat{k}$

$$= -48\hat{i} - 22\hat{j} + 8\hat{k}$$

$$|\vec{W}| = \sqrt{48^2 + 22^2 + 8^2} = 53.4$$

(ii)   As the required point is on z-axis so x = 0, y = 0

$$\vec{W} = -2z\hat{j} + 2z^2k^2 \text{ for that point.}$$

$\therefore$ $|\vec{W}| = \sqrt{(-2z)^2 + (2z^2)^2} = \sqrt{4z^2 + 4z^4} = 1$

so $4z^4 + 4z^2 - 1 = 0$

$$z^2 = \dfrac{-4 \pm \sqrt{16 - 16(-1)}}{8} = -\dfrac{1}{2} \pm \sqrt{\dfrac{1}{2}}$$

$$z^2 = -1.207 \text{ and } 0.207$$

taking z as positive $z^2 = 0.207$

$$z = \pm 0.455$$

Q.21: Calculate the differential volume to obtain the expression for volume of the

(i)   sphere of radius 'b'

(ii)   Semispherical shell of inner radius 'a' and outer radius 'b'.

(iii) Cylinder of radius 'b' and height 'h'.

Solution:

(i)   Differential volume in spherical coordinates

$$dV = r^2 \sin\theta \, dr \, d\theta \, d\phi$$

here $r = 0$ to b, $\theta = 0$ to π, $\phi = 0$ to 2π.

So volume of sphere

$$V = \int_V dV = \int_0^b \int_0^\pi \int_0^{2\pi} r^2 \sin\theta \, dr \, d\theta \, d\phi$$

$$= \int_0^b r^2 dr \int_0^\pi \sin\theta \, d\theta \int_0^{2\pi} d\phi$$

$$= 2\pi \int_0^b r^2 dr \left. (-\cos\theta) \right|_0^\pi$$

$$= 2\pi (1+1) \int_0^b r^2 dr$$

$$= 4\pi \frac{b^3}{3}$$

so $V_{sphere} = \dfrac{4\pi}{3} b^3$

(ii)   For semispherical shell

$$r_1 = a, \quad r_2 = b$$

so $dV = r^2 \sin\theta \, dr \, d\theta \, d\phi$

here $r = a$ to b, $\theta = 0$ to π, $\phi = 0$ to π

so $V = \displaystyle\int_a^b \int_0^\pi \int_0^\pi r^2 \sin\theta \, dr \, d\theta \, d\phi$

$$= \int\limits_{a}^{b} r^2 dr \int\limits_{0}^{\pi} \sin\theta \, d\theta \int\limits_{0}^{\pi} d\phi$$

$$= \pi \int\limits_{a}^{b} r^2 dr (\cos\theta) \Big|_{0}^{\pi}$$

$$= 2\pi \frac{r^2}{3} \Big|_{a}^{b}$$

$$= \frac{2\pi}{3} \left(b^3 - a^3 \right)$$

(iii)   Differential volume for a cylinder

$$dV = r \, dr \, d\phi \, dz$$

here   r = 0 to b,   z = 0 to h,   and ϕ = 0 to 2π.

so $$V = \int\limits_{V} dV$$

$$V = \int\limits_{a}^{b} \int\limits_{0}^{h} \int\limits_{0}^{2\pi} r \, dr \, d\phi \, dz$$

$$= \int\limits_{0}^{b} r \, dr \int\limits_{0}^{h} dz \int\limits_{0}^{2\pi} d\phi$$

$$= 2\pi . h . \frac{r^2}{2} \Big|_{0}^{b}$$

so $$V_{cyl.} = \pi b^2 h$$

Q.22 : **For positive x, y, z let ρ = 40 xyz c/m³. Find the total charge within the region bounded by x = 0, y = 0, $0 \le 2x + 3y \le 10$ and $0 \le z \le 2$.**

Solution : Here

$$Q = \int\limits_{0}^{5} \int\limits_{0}^{y} \int\limits_{0}^{z} 40 \, xyz \, dx \, dy \, dz$$

$$= \int_0^5 40x \; dx \left[\frac{y^2}{2}\right]_0^{\frac{10-2x}{3}} \left[\frac{z^2}{2}\right]_0^2$$

$$= \frac{40}{9} \int_0^5 \left(100x - 40x^2 + 4x^3\right) dx$$

$$= \frac{40}{9}\left[\frac{100x^2}{2} - \frac{40x^3}{3} - \frac{4x^4}{4}\right]_0^5$$

$$Q = 925.926 \text{ C}$$

Q.23: Given point P in Cartesian coordinate system as P(1, 2, 3). Calculate its coordinates in cylindrical system.

Solution: As given x = 1, y = 2, z = 3

$$\rho = \sqrt{x^2 + y^2} \quad \phi = \tan^{-1}\frac{y}{x}, \; z = z$$

so
$$\rho = \sqrt{1^2 + 2^2} = \sqrt{5} = 2.236$$

$$\phi = \tan^{-1}\frac{2}{1} = 63.43°$$

$$z = 3$$

so
$$P_{cyl.} = (2.236, 63.43°, 3)$$

Q.24: The cooridnate of a point P in cylindrical system is P(1, 45°, 2). find its equivalent in cartesion system.

Solution:

Here
$$\rho = 1, \quad \phi = 45°, \quad z = 2$$

and
$$x = \rho\cos\phi, \; y = \rho\sin\phi, \; z = z$$

$$x = 1.\cos 45 = \frac{1}{\sqrt{2}} = 0.707$$

$$y = 0.707$$

$$z = 2$$

so
$$P_{cart.} = (0.707, 0.707, 2)$$

Q.25: Find the constant m such that the vector

$$\vec{\phi} = (x+3y)\hat{i} + (y-2z)\hat{j} + (x+mz)\hat{k} \quad \text{is solenoidal.}$$

Solution: The vector will be solenoidal if $\vec{\nabla}.\vec{\phi} = 0$

so $\left(\dfrac{\partial}{\partial x}\hat{i} + \dfrac{\partial}{\partial y}\hat{j} + \dfrac{\partial}{\partial z}\hat{k}\right)\left[(x+3y)\hat{i} + (y-2z)\hat{j} + (x+mz)\hat{k}\right] = 0$

i.e. $\dfrac{\partial}{\partial x}(x+3y) + \dfrac{\partial}{\partial y}(y-2z) + \dfrac{\partial}{\partial z}(x+mz) = 0$

or $\qquad\qquad\qquad 1 + 1 + m = 0$

$$m = -2$$

Q.26: Find div $\vec{F}$ and curl $\vec{F}$ if $\vec{F}$ = grad $\left(x^3 + y^3 + z^3 - 3xyz\right)$.   **[WBUT 2001]**

Solution: Here $\qquad\qquad \vec{F} = \text{grad}\left(x^3 + y^3 + z^3 - 3xyz\right)$

$$= \dfrac{\partial}{\partial x}\left(x^3 + y^3 + z^3 - 3xyz\right)\hat{i} + \dfrac{\partial}{\partial y}\left(x^3 + y^3 + z^3 - 3xyz\right)\hat{j}$$

$$+ \dfrac{\partial}{\partial z}\left(x^3 + y^3 + z^3 - 3xyz\right)\hat{k}$$

$$= \left(3x^2 - 3yz\right)\hat{i} + \left(3y^2 - 3xz\right)\hat{j} + \left(3z^2 - 3xy\right)\hat{k}$$

$$\text{div } \vec{F} = \dfrac{\partial}{\partial x}\left(3x^2 - 3yz\right) + \dfrac{\partial}{\partial y}\left(3y^2 - 3xz\right) + \dfrac{\partial}{\partial z}\left(3z^2 - 3xy\right)$$

$$= 6(x+y+z)$$

$$\vec{\nabla} \times \vec{F} = \begin{vmatrix} \hat{i} & \hat{j} & \hat{k} \\ \dfrac{\partial}{\partial x} & \dfrac{\partial}{\partial y} & \dfrac{\partial}{\partial z} \\ 3x^2 - 3yz & 3y^2 - 3xz & 3z^2 - 3xy \end{vmatrix}$$

$$= \left(-3x + 3x\right)\hat{i} + \left(-3y + 3y\right)\hat{j} + \left(-3z + 3z\right)\hat{k} = 0$$

Q.27 Show that curl grad f = 0 where f = $x^2y + 2xy + z^2$.

Solution:

$$\text{grad } f = \frac{\partial f}{\partial x}\hat{i} + \frac{\partial f}{\partial y}\hat{j} + \frac{\partial f}{\partial z}\hat{k}$$

$$= (2xy + 2y)\hat{i} + (x^2 + 2x)\hat{j} + (2z)\hat{k}$$

$\therefore$ $$\text{curl grad } f = \begin{vmatrix} \hat{i} & \hat{j} & \hat{k} \\ \dfrac{\partial}{\partial x} & \dfrac{\partial}{\partial y} & \dfrac{\partial}{\partial z} \\ 2xy+2y & x^2+2x & 2z \end{vmatrix}$$

$$= 0 + 0 + (2x + 2 - 2x - 2)\,\hat{k} = 0$$

Q.28: If the scalar function $\psi(x,y,z) = 2xy + z^2$,. is its corresponding scalar field is solenoidal or irrotational?

Solution : Let

$$\vec{F} = \nabla\psi = 2y\hat{i} + 2x\hat{j} + 2z\hat{k}$$

so

$$\vec{\nabla}.\vec{F} = \frac{\partial}{\partial x}(2y) + \frac{\partial}{\partial y}(2x) + \frac{\partial}{\partial z}(2z)$$

$$= 0 + 0 + 2 = 2$$

$$\neq 0$$

So field is not solenoidal.

Now

$$\vec{\nabla} \times \vec{F} = \begin{vmatrix} \hat{i} & \hat{j} & \hat{k} \\ \dfrac{\partial}{\partial x} & \dfrac{\partial}{\partial y} & \dfrac{\partial}{\partial z} \\ 2y & 2x & 2z \end{vmatrix}$$

$$= 0$$

so field is irrotational.

Q.29: Verify the divergence theorem for the vector function

$$\vec{F} = 4xz\hat{i} - y^2\hat{j} + yz\hat{k}$$

taken over the cube bounded by x = 0, 1 y = 0, 1, z = 0, 1.

[WBUT (math) 2002]

Solution.:

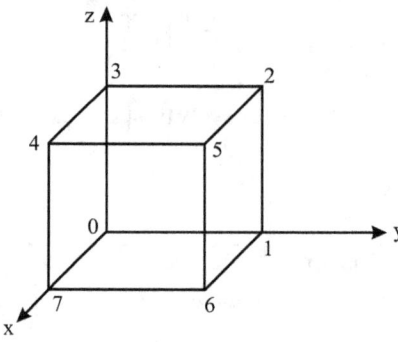

for face 4567; $\hat{n} = \hat{i}$ and x = 1

$$\oint_1 \vec{F} . \hat{n}\ ds \ = \ \int\limits_0^1 \int\limits_0^1 4z\ dy\,dz = 2$$

$$\oint_2 \vec{F} . \hat{n}\ ds \ = \ \left(-\hat{i}\right)dy\ dz = 0 \ \ \text{for } 1230$$

for $2561\ \hat{n} \ = \ \hat{j},\ y = 1$

$$\oint_3 \vec{F} . \hat{n}\ ds \ = -1$$

and for 3074 $$\oint_4 \vec{F} . \hat{n}\ ds \ = 0$$

for face 2345, $$\oint_5 \vec{F} . \hat{n}\ ds \ = \frac{1}{2}$$

and $$\oint_6 \vec{F} . \hat{n}\ ds \ = 0 \ \text{for } 0167$$

total $$\oint_S \vec{F} . \hat{n}\ ds \ = 2 + 0 + (-1) + 0 + \frac{1}{2} + 0 = \frac{3}{2}$$

Again $$\vec{\nabla} . \vec{F} \ = \ \frac{\partial}{\partial x}\left(4xz\right) + \frac{\partial}{\partial y}\left(-y^2\right) + \frac{\partial}{\partial z}\left(yz\right)$$

$$= 4z - y$$

$$\therefore \qquad \int_0^1 \int_0^1 \int_0^1 \vec{\nabla}.\vec{F}\,dV = \int_0^1 \int_0^1 \int_0^1 (4z - y)dx\,dy\,dz$$

$$= \int_0^1 \int_0^1 \left[\frac{4z^2}{2} - yz\right]_0^1 dx\,dy$$

$$= \frac{3}{2}$$

$$\therefore \qquad \oint_s \vec{F}.\hat{n}\,ds = \oint_V \vec{\nabla}.\vec{F}\,dV$$

Hence divergence theorem is verified.

Q.30: Calculate the line integral of $\vec{A} = \rho\cos\phi\,\hat{\rho} + z\sin\phi\,\hat{z}$ around the edge L of the wedge defined by $0 \le \rho \le 4$, $0 \le \phi \le 30°$, $z = 0$.

Solution:

Given
$$\vec{A} = \rho\cos\phi\,\hat{\rho} + z\sin\phi\,\hat{z}$$

differential length
$$\vec{dl} = d\rho\,\hat{\rho} + d\phi\,\hat{\phi} + dz\,\hat{z}$$

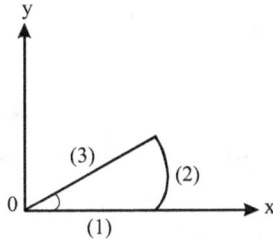

Circulation of $\vec{A}$ around path is

$$\oint_L \vec{A}.\vec{dl} = \oint_1 \vec{A}.\vec{dl} + \oint_2 \vec{A}.\vec{dl} + \oint_3 \vec{A}.\vec{dl}$$

for (1) $\phi = 0$, $d\phi = 0$, $z = 0$, $dz = 0$

$$\oint_1 \vec{A}.\vec{dl} = \oint_1 (\rho\cos\phi\,\hat{\rho} + z\sin\phi)\cdot(d\rho\,\hat{\rho} + d\phi\,\hat{\phi} + dz\,\hat{z})$$

$$= \oint_1 \rho\cos\phi\,d\rho = \int_0^4 \rho\,d\rho = \frac{\rho^2}{2}\Big|_0^4$$

$$= \frac{16}{2} = 8$$

for (2) $\rho = 4,\ d\rho = 0$

$z = 0,\ dz = 0$

$$\oint_2 \vec{A}.\,\overrightarrow{dl} = \oint_2 \rho \cos \phi\, d\rho + z \sin \phi\, dz = 0$$

for (3) $\phi = \pi/6,\ d\phi = 0,\ \ z = 0,\ \ \ dz = 0$

$$\oint_3 \vec{A}.\,\overrightarrow{dl} = \oint_3 \rho \cos \phi\, d\rho$$

$$= \int_4^0 \rho \cos \frac{\pi}{6}\, d\rho$$

$$= \frac{\sqrt{3}}{2} \int_4^0 \rho\, d\rho = 0.866 \left.\frac{\rho^2}{2}\right|_4^0$$

$$= -6.93$$

So total $\oint_L \vec{A}.\,\overrightarrow{dl} = 8 + 0 - 6.93$

$$= 1.07$$

Q.31 Given $\vec{A} = x^2 + xy$, **calculate** $\int \vec{A}.\,\overrightarrow{ds}$ **over the region** $y = x^2$, $0 < x < 2$.

Solution:

So $y = x^2,\ \ \ \ ds = dx\, dy$

$$\int \vec{A}.\,\overrightarrow{ds} = \int \left(x^2 + xy\right) dx\, dy$$

$$= \int_{0}^{2}\int_{y=0}^{x^2}\left(x^2 \, dx \, dy\right) + \int_{0}^{2}\int_{y=0}^{x^2} xy \, dx \, dy$$

$$= \int_{x=0}^{2} x^4 dx + \int_{x=0}^{2} \frac{x^5}{2} \, dx$$

$$= \frac{32}{5} + \frac{64}{12}$$

$$= 11.7$$

Q.32 For a scalar function $\phi = \left[\sin\dfrac{\pi x}{2} + \sin\dfrac{\pi y}{3}\right] e^{-z}$. Calculate the magnitude direction of maximum rate of increase of ϕ at the point (1, 1, 1).

Solution: As gradient of a scalar function gives the magnitude and direction of max. rate of change of that

So
$$\nabla\phi = \frac{\partial\phi}{\partial x}\hat{i} + \frac{\partial\phi}{\partial y}\hat{j} + \frac{\partial\phi}{\partial z}\hat{k}$$

$$= \frac{\partial}{\partial x}\left(\sin\frac{\pi x}{2}\sin\frac{\pi y}{3}e^{-z}\right)\hat{i} + \frac{\partial}{\partial y}\left(\sin\frac{\pi x}{2}\sin\frac{\pi y}{3}e^{-z}\right)\hat{j}$$

$$+ \frac{\partial}{\partial z}\left(\sin\frac{\pi x}{2}\sin\frac{\pi y}{3}e^{-z}\right)\hat{k}$$

$$= \left(\frac{\pi}{6}e^{-z}\sin\frac{\pi x}{2}\right)\hat{j} - \left(\sin\frac{\pi x}{2}\sin\frac{\pi y}{3}e^{-z}\right)\hat{k}$$

at (1, 1, 1)

$$(\nabla\phi)_{1,1,1} = \left(\frac{\pi}{6}e^{-1}\sin\frac{\pi}{2}\right)\hat{j} - \left(\sin\frac{\pi}{2}\sin\frac{\pi}{3}e^{-1}\right)\hat{k}$$

$$= \left(\frac{\pi}{6}\times 0.367\right)\hat{j} - \left(0.866\times 0.367\right)\hat{k}$$

$$= 0.192\,\hat{j} - 0.318\,\hat{k}$$

and
$$|\nabla\phi|_{1,1,1} = \left[(0.192)^2 + (0.318)^2\right]^{1/2}$$

$$= 0.37$$

and
$$\hat{j} = \frac{\nabla\phi}{|\nabla\phi|} = \frac{0.192\hat{j} - 0.318\hat{k}}{0.37}$$

$$= 0.52\hat{j} - 0.86\hat{k}$$

Q.33 : Determine the divergence of the following vector fields at given points–

(i) $\vec{A} = yz\hat{i} + 4(x+y)\hat{j} + xyz\,\hat{k}$ at $(1,-2,1)$

(ii) $\vec{B} = \rho z \sin\phi\hat{\rho} + 5\rho z \cos\phi\,\hat{\phi} + z\hat{z}$ at $\left[5, \frac{\pi}{2}, 1\right]$

(iii) $\vec{C} = 2r\sin\theta\cos\phi\,\hat{r} + \cos\phi\,\hat{\theta} + r\,\hat{\phi}$ at $\left[1, \frac{\pi}{3}, \frac{\pi}{3}\right]$

Solution: In cartesion

(a)
$$\vec{\nabla}.\vec{A} = \frac{\partial}{\partial x}(yz) + \frac{\partial}{\partial y}(4x+4y) + \frac{\partial}{\partial z}(xyz)$$

$$= 0 + 4 + xy$$

$$\vec{\nabla}.\vec{A} = 4 + xy$$

at $(1,-2,1)$, $\left(\vec{\nabla}.\vec{A}\right)_{1,-2,1} = 4 + 1 \times (-2)$

$$= 2$$

(b)   In cylindrical

$$\vec{\nabla}.\vec{B} = \frac{1}{\rho}\frac{\partial}{\partial\rho}\left(\rho\, B_\rho\right) + \frac{1}{\rho}\frac{\partial}{\partial\phi}\left(B_\phi\right) + \frac{\partial}{\partial z}B_z$$

given
$$B_\rho = \rho z \sin\phi_1 \quad B_\phi = 5\rho z \cos\phi, \ B_z = z$$

$$\vec{\nabla}.\vec{B} = \frac{2}{\rho}\rho z \sin\phi + \frac{1}{\rho}5.\rho z(-\sin\phi) + 1$$

$$= 1 - 3z\sin\phi$$

at $\left(5, \frac{\pi}{2}, 1\right) \Rightarrow$

$$\vec{\nabla}.\vec{B} = 1 - (3 \times 1 \times 1)$$

$$= -2$$

(c)   In spherical system

$$\vec{\nabla}.\vec{C} = \frac{1}{r^2}\frac{\partial}{\partial r}\left(r^2 C_r\right) + \frac{1}{r\sin\theta}\frac{\partial}{\partial\theta}\left(\sin\theta\, C_\theta\right) + \frac{1}{r\sin\theta}\frac{\partial C_\phi}{\partial\phi}$$

$$= \frac{2}{r^2}\sin\theta\cos\phi\frac{\partial}{\partial r}\left(r^3\right) + \frac{\cos\phi}{r\sin\theta}\frac{\partial}{\partial\theta}\sin\theta\frac{\partial}{r\sin\theta}\frac{\partial}{\partial\phi}$$

$$= 6\sin\theta\cos\phi + \frac{\cos\phi\cot\theta}{r}$$

at $\left(1, \dfrac{\pi}{3}, \dfrac{\pi}{3}\right)$

$$\vec{\nabla}.\vec{C} = 6\sin\frac{\pi}{3}\cos\frac{\pi}{3} + \frac{\cos\dfrac{\pi}{3}\cot\dfrac{\pi}{3}}{1}$$

$$= 2.6 + 0.288$$

$$= 2.88$$

Q.34: Find the nature of the vector $\vec{F} = 30\hat{i} + 2xy\,\hat{j} + 5xz^2\hat{k}$.   [RU 2003]

Solution :

$$\vec{\nabla}.\vec{F} = \frac{\partial}{\partial x}(30) + \frac{\partial}{\partial y}(2xy) + \frac{\partial}{\partial y}\left(5xz^2\right)$$

$$= 2x + 10xz$$

$$\neq 0$$

$$\text{Curl }\vec{F} = \vec{\nabla}\times\vec{F} = \begin{vmatrix} \hat{i} & \hat{j} & \hat{k} \\ \dfrac{\partial}{\partial x} & \dfrac{\partial}{\partial y} & \dfrac{\partial}{\partial z} \\ 30 & 2xy & 5xz^2 \end{vmatrix}$$

$$= 5z^2\,\hat{j} + 2y\hat{k}$$

$$\neq 0$$

as $\vec{\nabla}.\vec{F} \neq 0 \Rightarrow$ fields is not solenoidal

and $\vec{\nabla}\times\vec{F} \neq 0 \Rightarrow$ field is rotational.

Q.35: Given the vector field $\vec{G} = (16xy - 3)\hat{i} + 8x^2\hat{j} - x\hat{k}$.

(i)   Is G irrotational (or conservative)?

(ii)   Find the net flux of G over the cube $0 < x, y, z < 1$.

(iii) Determine the circulation of G around the edge of the square $z = 0$, $0 < x$, $y < 1$. Assume anticlockwise direction.   [RU 2003]

Solution:

(i)
$$\vec{\nabla} \times \vec{G} = \begin{vmatrix} \hat{i} & \hat{j} & \hat{k} \\ \dfrac{\partial}{\partial x} & \dfrac{\partial}{\partial y} & \dfrac{\partial}{\partial z} \\ 16xy - z & 8x^2 & -x \end{vmatrix}$$

$$= 0\hat{i} + (-1 + 1)\hat{j} + (16x - 16x)\hat{k} = 0$$

So $\vec{G}$ is irrotational.

(ii)   Net flux of $\vec{G}$ over the cube

$$= \int_V (\vec{\nabla} \cdot \vec{G}) dV$$

$$\vec{\nabla} \cdot \vec{G} = \frac{\partial}{\partial x}(16xy - z) + \frac{\partial}{\partial y}(8x^2) + \frac{\partial}{\partial z}(-x)$$

$$= 16y + 0 + 0 = 16y$$

so
$$\int_V (\vec{\nabla} \cdot \vec{G}) dV = \iiint 16y \, dx \, dy \, dz = 16 \int_0^1 dx \int_0^1 dz \int_0^1 y \, dy$$

$$= 16 \cdot 1 \cdot 1 \left.\frac{y^2}{2}\right|_0^1 = 8$$

(iii)
$$\int_0^1 \vec{G} \cdot \vec{dl} = \int_{x=0}^1 (16xy - z)dx \left|_{z=0}^{y=0}\right. + \int_{y=0}^1 8x^2 dy \left|_{z=0}^{x=1}\right.$$

$$+ \int_{x=1}^0 (16xy - z)dx \left|_{z=0}^{y=1}\right. + \int_{y=1}^0 8x^2 dy \left|_{z=0}^{x=0}\right.$$

$$= 0 + 8(1)y|_0^1 + 16(1)\left.\frac{x^2}{2}\right|_1^0 + 0$$

$$= 8 - 8 = 0.$$

SUMMARY

- A vector is with magnitude and direction.
- In space a quantity is specified by a function.
- When the result of multiplication of two vectors is a scalar then it is called scalar product or dot product.
- When product is a vector then it is called vector product.
- Multiplication of three vectors can give scalar $\vec{A} \cdot (\vec{B} \times \vec{C})$ or a vector $\vec{A} \times (\vec{B} \times \vec{C})$.
- Vector differentiation is done using dal (∇) operator the gradient of a scalar field is $\nabla \phi$, divergence as $\vec{\nabla} \cdot \vec{A}$ and curl by $\vec{\nabla} \times \vec{A}$ and laplacian by $\nabla^2 A$.
- In Cartesion coordinate system

$$\vec{dl} = dx\,\hat{i} + dy\,\hat{j} + dz\,\hat{k}, \quad dV = dx\,dy\,dz$$

$$\text{Gradient } \nabla\phi = \frac{\partial\phi}{\partial x}\hat{i} + \frac{\partial\phi}{\partial y}\hat{j} + \frac{\partial\phi}{\partial z}\hat{k}$$

$$\text{Divergences } \vec{\nabla} \cdot \vec{A} = \frac{\partial A_x}{\partial x} + \frac{\partial A_y}{\partial y} + \frac{\partial A_z}{\partial z}$$

$$\text{Curl } \vec{\nabla} \times \vec{A} = \begin{vmatrix} \hat{i} & \hat{j} & \hat{k} \\ \frac{\partial}{\partial x} & \frac{\partial}{\partial x} & \frac{\partial}{\partial z} \\ A_x & A_y & A_z \end{vmatrix}$$

$$\text{Laplacian } \nabla^2\phi = \frac{\partial^2\phi}{\partial x^2} + \frac{\partial^2\phi}{\partial y^2} + \frac{\partial^2\phi}{\partial z^2}$$

- In cylindrical system

$$\vec{dl} = d\rho\,\hat{\rho} + \rho\,d\phi\,\hat{\phi} + dz\,\hat{z}, \quad dV = \rho\,d\phi\,d\rho\,dz$$

$$\text{gradient } \nabla T = \left[\frac{\partial T}{\partial \rho}\hat{\rho} + \frac{1}{\rho}\frac{\partial T}{\partial \phi}\hat{\phi} + \frac{\partial T}{\partial z}\hat{z}\right]$$

$$\text{divergence } \vec{\nabla} \cdot \vec{A} = \frac{1}{\rho}\frac{\partial}{\partial \rho}\left(\rho A_\rho\right) + \frac{1}{\rho}\frac{\partial A_\phi}{\partial \phi} + \frac{\partial A_z}{\partial z}$$

$$\text{Curl } \vec{\nabla} \times \vec{A} = \frac{1}{\rho} \begin{vmatrix} \hat{\rho} & \hat{\phi} & \hat{z} \\ \frac{\partial}{\partial \rho} & \frac{\partial}{\partial \phi} & \frac{\partial}{\partial z} \\ A_\rho & \rho A_\phi & A_z \end{vmatrix}$$

$$\text{Laplacian } \nabla^2 \phi = \frac{1}{\rho} \frac{\partial}{\partial \rho} \left(\rho \frac{\partial \phi}{\partial \rho} \right) + \frac{1}{\rho^2} \frac{\partial^2 \phi}{\partial \phi^2} + \frac{\partial^2 \phi}{\partial z^2}$$

- In spherical system

$$\vec{dl} = dr\,\hat{r} + rd\theta\,\hat{\theta} + r\sin\theta\,d\phi\,\hat{\phi}$$

$$dV = r^2 \sin\theta\, d\theta\, dr\, d\phi$$

$$\text{gradient } \nabla T = \frac{\partial T}{\partial r}\hat{r} + \frac{1}{r}\frac{\partial T}{\partial \theta}\hat{\theta} + \frac{1}{r\sin\theta}\frac{\partial T}{\partial \phi}\hat{\phi}$$

$$\text{divergence } \vec{\nabla}.\vec{A} = \frac{1}{r^2}\frac{\partial}{\partial r}\left(r^2 A_r\right) + \frac{1}{r\sin\theta}\frac{\partial}{\partial \theta}\left(A_\theta \sin\theta\right) + \frac{1}{r\sin\theta}\frac{\partial A_\phi}{\partial \phi}$$

$$\text{Curl } \vec{\nabla} \times \vec{A} = \frac{1}{r\sin\theta} \begin{vmatrix} \hat{r} & \hat{\theta} & r\sin\theta\,\hat{\phi} \\ \frac{\partial}{\partial r} & \frac{\partial}{\partial \theta} & \frac{\partial}{\partial \phi} \\ A_r & rA_\theta & r\sin\theta\,A_\phi \end{vmatrix}$$

$$\text{Laplacian } \nabla^2 \phi = \frac{1}{\rho}\frac{\partial}{\partial r}\left(r^2 \frac{\partial \phi}{\partial r}\right) + \frac{1}{r^2 \sin\theta}\frac{\partial}{\partial \theta}\left(\sin\theta \frac{\partial \phi}{\partial \theta}\right)$$

$$+ \frac{1}{\partial^2 \sin^2\theta}\frac{\partial^2 \phi}{\partial \phi^2}$$

- Gauss Divergence theorem

$$\oint_s \vec{A}.\vec{ds} = \int_V \left(\vec{\nabla}.\vec{A}\right) dV$$

- Stoke's theorem

$$\oint_L \vec{A}.\vec{dl} = \int_V \left(\vec{\nabla} \times \vec{A}\right).\vec{ds}$$

- A vector field is solenoidal if $\vec{\nabla} \cdot \vec{A} = 0$

  Irrotational or conservative if $\vec{\nabla} \times \vec{A} = 0$

- Triple products

  $$\vec{A} \cdot \left(\vec{B} \times \vec{C} \right) = \vec{B} \left(\vec{C} \times \vec{A} \right) = \vec{C} \left(\vec{A} \times \vec{B} \right)$$

  $$\vec{A} \times \left(\vec{B} \times \vec{C} \right) = \vec{B} \left(\vec{A} \cdot \vec{C} \right) - \vec{C} \left(\vec{A} \cdot \vec{B} \right)$$

- Second derivatives

  $$\vec{\nabla} \cdot \left(\vec{\nabla} \times \vec{A} \right) = 0$$

  $$\vec{\nabla} \times \left(\vec{\nabla} f \right) = 0$$

  $$\vec{\nabla} \times \left(\vec{\nabla} \times \vec{A} \right) = \vec{\nabla} \left(\vec{\nabla} \cdot \vec{A} \right) - \vec{\nabla}^2 \vec{A}$$

EXERCISE

1. Give the basic concepts of transformation of one coordinate system to another. Derive necessary relations for rectangular, cylindrical and spherical systems. **[RU 2003]**

2. Write short note on "Physical significance of curl, divergence and gradient".

3. State-Gauss divergence theorem. Write its applications, advantages and limitations. **[RU 2002]**

4. State and prove stoke's theorem. **[RU 2000]**

5. Explain how stoke's theorem enables us to obtain the integral form of ampere circuital law.

6. Explain various types of vector fields.

    (i)   Solenoidal and irrotational fields.

    (ii)   Irrotational but not solenoidal fields

    (iii) Solenoidal but not irrotaitonal fields

    (iv) Neither irrotational nor solenoidal fields.

7. For the vectors $\vec{A} = \hat{i} + 3\hat{k}$ and $\vec{B} = 5\hat{i} + 2\hat{j} - 6\hat{k}$ calculate

    (i) $\vec{A} + \vec{B}$ (ii) $\vec{A} - \vec{B}$

    (iii) $\vec{A} \cdot \vec{B}$ (iv) $\vec{A} \times \vec{B}$

    (v)   Angle between $\vec{A}$ and $\vec{B}$

(vi)   A unit vector parallel to $3\vec{A} + \vec{B}$.

(vii) Length of the projection of $\vec{A}$ on $\vec{B}$.

Ans.: $(\mathbf{i})\ 6\hat{i} + 2\hat{j} - 3\hat{k}$ $\qquad (\mathbf{ii})\ -4\hat{i} - 2\hat{j} + 9\hat{k}$ $\qquad (\mathbf{iii})-13$ $\qquad (\mathbf{iv})-6\hat{i} + 21\hat{j} + 2k$

$\qquad (\mathbf{v})\ 60°$ $\qquad\qquad (\mathbf{vi})\ \dfrac{8\hat{i} + 2\hat{j} + 3\hat{k}}{\sqrt{77}}$ $\qquad (\mathbf{vii})\ 1.6m$

8.   Use the differential volume dV to find volume of region.

(i) $0 < x < 1,$ $\qquad 1 < y < 2,$ $\qquad -3 < z < 3$ $\hfill$ [Ans.: **6**]

(ii) $2 < \rho < 5,$ $\qquad \dfrac{\pi}{3} < \phi < \pi,$ $\qquad -1 < z < 4$ $\hfill$ [Ans.: **110**]

9.   Find area of the region $0 \le \phi \le \beta$ on the spherical shell of radius 'b'. $\hfill$ [Ans. $2\beta b^2$]

10. Evaluate the gradient of the following scalar fields

(a) $P = e^{-z} \sin 2x$ $\hfill$ [Ans.: $\nabla P = 2\cos 2x\,e^{-z}\hat{i} - \sin 2x\,e^{-z}\hat{k}$]

(b) $q = \rho^2 z \cos \phi$ $\hfill \left[\text{Ans.: } \nabla q = 2\rho z \cos\phi\,\hat{\rho} - \rho^2 z \sin\phi\,\hat{\phi} + \rho^2 \cos\phi z\right]$

(c) $s = 20r \sin\theta \cos\phi$ $\hfill \left[\begin{array}{l}\text{Ans.: } \nabla r = 20\sin\theta\cos\phi\hat{r} + 20r\cos\phi \\ \qquad\qquad \cos\theta\,\hat{\theta} - 20r\sin\theta\,\sin\phi\,\hat{\phi}\end{array}\right]$

11. If $f = xy + yz + xz$ then

(i)   Find the magnitude and direction of the maximum rate of change of the function at point (1, 2, 3)

(ii)   Find the rate of change of the function at the same point in the direction of the vector.

$$\left[\begin{array}{l} \text{Ans.:}(\mathbf{i})\ |\nabla f| = \sqrt{14} \\[4pt] \qquad \left(\hat{\nabla}f\right) = \dfrac{\nabla f}{|\nabla f|} = \dfrac{2\hat{i} + 3\hat{j} + \hat{k}}{\sqrt{14}} \\[4pt] (\mathbf{ii})\ df = \nabla f . dl = 11 \end{array}\right]$$

12. If $T = 2x\hat{i} + 3y\,\hat{j} - 4z\hat{k}$ and $V = xyz$ evaluate $\vec{\nabla}.\left(V\vec{T}\right)$. $\hfill$ [Ans.: **2xyz**]

13. If $U = xz - x^2 y + y^2 z^2$ find div (grad U). $\hfill$ [Ans.: $2(-y + z^2 + y^2)$]

14. Given $\vec{D} = 6xyz^2\,\hat{i} + 3x^2z^2\,\hat{j} + 6x^2y\,\hat{k}$ C/m^3 Find the total charge lying within the region bounded by $0 < x < 1$, $1 < y < 2$ and $|z| \le 1$ by separately evaluating each side of divergence theorem. **[Ans.: 6C]**

15. If $\vec{A} = (x+y+1)\hat{i} + \hat{j} - (x+y)\hat{k}$ then prove that $\vec{A} \cdot (\vec{\nabla} \times \vec{A}) = 0$

16. Prove that vector $\vec{A} = (x - yz)\hat{i} + (2y - zx)\hat{j} + (2z - xy)\hat{k}$ is not solenoidal. **[WBUT 2005]**

17. If $\phi = x^2 - y^2 + 2z$ find $\vec{\nabla} \cdot (\vec{\nabla}\,\phi)$. **[WBUT 2005]**

18. Show that $\vec{B} = 2xyz\,\hat{i} + (x^2z + 2y)\,\hat{j} + x^2y\hat{k}$ is irrotational. **[WBUT 2005]**

19. find a unit vector perpendicular to $x^2 + y^2 - z^2 = 100$ at (1, 2, 3). **[WBUT 2007]**

$$\left[\text{Ans.}: \frac{\hat{i} + 2\hat{j} - 3\hat{k}}{\sqrt{14}}\right]$$

20. if $\phi = 3x^2y - y^3z^2$ find $\vec{\nabla}\phi$ at (1, –2, –1). **[WBUT 2004]**

$$\left[\text{Ans.}: \vec{\nabla}\phi = -12\hat{i} - 9\hat{j} - 16\hat{k}\right]$$

21. Show that $\vec{F} = (2xy + z^3)\hat{i} + x^2\hat{j} + 3xz^2\hat{k}$ is a conservative force field. Find also the scalar potential. **[WBUT 2006, 2003]** $\left[\text{Ans.}: \phi = x^2y + z^3x + \text{constant}\right]$

22. Evaluates $\oint_s \vec{F} \cdot \hat{n}\, ds$ where $\vec{F} = 8xz\hat{i} - y^2\hat{j} + yz\hat{k}$ and s is the surface of the cube bounded by x = 0, 1, y = 0, 1, z = 0, 1.

2

ELECTRICITY

ELECTRICITY

Coulomb's law in vector form, Electrostatic field and its curt. Gauss's Law in integral form and conversion to differential form. Electrostatic potential and field. Poisson's equation, Laplace equation. (Application to Cartesian, spherical and cylindrically symmetric systems-- 1-D problem). Electric current, drift velocity, current density, continuity equation, steady current.

DIELECTRICS

Concepts of polarization, the relation $D = \varepsilon_0 E + P$, polarizability, electronic polarisation and polarisation in monoatomic and polyatomic gases.

Electrostatics

The fundamental physical quantity which constitute all electromagnetic fields in charge. Its measuring unit is coulomb (C) and characterized by electronic charge where one electron possesses a charge of 1.6×10^{-19} C (i.e. one coulomb of charge is represented by 6×10^{18} electrons). Electronic charge is assumed to be of –ve nature whereas +ve charge is associated with protons in the nucleus.

The complete study of different physical phenomenon between two or more charges when charges are at rest, is known as electrostatics.

When at rest, two charges placed near to each other can experience a force (attractive or repulsive) named as electrostatic forces which can be measured using coulomb's law of electrostatics. The nature of this force is dependent over the types of charges. If the charges are alike, the force will be repulsive otherwise forces will be attractive.

Electrostatic force - Coulomb's law

Colonel Chargles Augustin De Coulomb in 1785, gave a relation between the magnitude of the force exerted by a point charge q_1 by another charge q_2 when they are separated by a distances r in free spaces.

It states that the electrostatic force between two charges q_1 and q_2 is directly proportional to the product of their charges and inversely proportional to the square of the distance between them and acts along the joining line of two charges.

$$\vec{F} = k \frac{q_1 q_2}{r^2} \hat{r}$$

where   k = Coulomb constant

for free space $k = \dfrac{1}{4\pi\varepsilon_0}$

ε_0 = permittivity of free space

$$= 8.854 \times 10^{-12} \text{ Farad/meter.} \qquad \left(\text{or } \frac{(\text{Coulomb})^2}{\text{Newton }(\text{meter})^2} \right)$$

so that $k = 9 \times 10^9$ m/F

for any other medium

$$k = \frac{1}{4\pi\,\varepsilon_m}$$

where ε_m = permittivity in the medium

$= \varepsilon_r \varepsilon_0$ where ε_r is relative permittivity of the medium.

Vector form

If two point charges q_1 and q_2 are at a distance $\vec{r_1}$ and $\vec{r_2}$ from the origin, then

$$\vec{r}_{12} = \vec{r_2} - \vec{r_1}$$

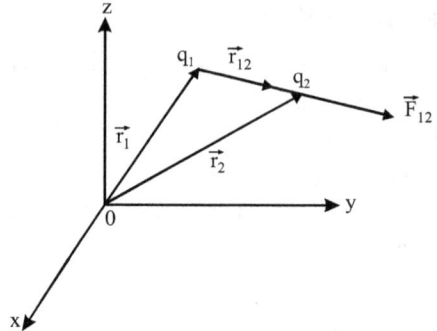

Fig.

so the force $\vec{F}_{12}$ acting on q_2 at $\vec{r}_2$ due to q_1 will be

$$\vec{F}_{12} = \frac{1}{4\pi\varepsilon_0} \frac{q_1 q_2}{r_{12}^2} \hat{r}_{12} = \frac{1}{4\pi\varepsilon_0} \frac{q_1 q_2}{r_{12}^3} \vec{r}_{12} \qquad \ldots(1)$$

Now

$$\vec{r}_{21} = \vec{r}_1 - \vec{r}_2 = -\vec{r}_{12}$$

and

$$\vec{F}_{21} = \frac{1}{4\pi\varepsilon_0} \frac{q_1 q_2}{r_{21}^2} \hat{r}_{21}$$

$$= \frac{1}{4\pi\varepsilon_0} \frac{q_1 q_2}{r_{21}^3} \hat{r}_{21}$$

$$= \frac{1}{4\pi\varepsilon_0} \frac{q_1 q_2}{r_{21}^3} \left(-\vec{r}_{12}\right) \quad \text{as} \ \vec{r}_{12} = -\vec{r}_{21} \qquad \ldots(2)$$

and

$$\left|\vec{r}_{12}\right| = \left|\vec{r}_{21}\right| = r_{12} = r_{21}$$

comparing equation (1) and (2)

$$= -\vec{F}_{12}$$

CHARGE DISTRIBUTIONS

(1)   Linear charge distribution

When a charge Δq is distributed over a conductor of infinitesimally small thickness along a length Δl then charge distribution is linear and linear charge density

$$\lambda = \lim_{\Delta l \to 0} \frac{\Delta q}{\Delta l} \ C/m$$

so that the total charge over the entire conductor of length l is

$$q = \int_0^1 \lambda \, dl$$

(2) Surface charge Distribution :

When some charge is distributed over a small area then surface charge density will be

$$\sigma = \lim_{\Delta s \to 0} \frac{\Delta q}{\Delta s} \; C/m^2$$

so total charge over the entire surface

$$q = \oint_s \sigma \, ds$$

(3) Volume charge distribution

When charge is distributed over some volume of a conducting medium then volume charge density can be defined as

$$\rho = \lim_{\Delta V \to 0} \frac{\Delta q}{\Delta V} \; C/m^3$$

where ΔV is infinitesimally small volume element so total charge within the volume

$$q = \int_V \rho \, dV$$

Electric Field

When a charge is placed near to another charge, it experiences an electrostatic force according to coulomb's law. The space around any charge upto which any other charge experiences a force, is called electric field or electrostatic field due to that charge.

Mathematically $\qquad \vec{E} = \dfrac{\vec{F}}{q} \; N/C \qquad \qquad$ ...(1)

If charge q_1 is placed near to charge q_2 then electric field due to q_1 is

$$\vec{E} = \frac{1}{4\pi\varepsilon_0} \frac{q_1 q_2}{r^2} \frac{1}{q_2}$$

$$\vec{E} = \frac{1}{4\pi\varepsilon_0} \frac{q_1}{r^2} \hat{r} \frac{N}{C} \qquad \qquad \text{...(2)}$$

Electric field intensity $\left(\vec{E}\right)$ or the strength of an electric field at a given point is the force exerted on a unit +ve charge placed at that point.

Here it can be summed up that the electrostatic force experienced by a charge q_2 in an electric field $\vec{E}$ is

$$\vec{F} = q_2 \vec{E} \text{ N} \qquad \qquad ...(3)$$

Important Point (comparison of electrostatic forces with Gravitational force)

1. Electrostatic force can be attractive or repulsive but gravitational force can only be attractive in nature.

2. Charges can be +ve or –ve but mass is always +ve in nature.

3. Coulombs law of electrostatic force is also a inverse square law similar to Newton's law of Gravitation.

Electric Displacement Vector

The electric displacement vector $\left(\vec{D}\right)$ at any point in an isotropic linear medium having electrical permittivity ε_m is related with electric field intensity $\left(\vec{E}\right)$ is given by

$$\vec{D} = \varepsilon_m \vec{E}$$

In free space

$$\vec{D}_0 = \varepsilon_0 \vec{E}$$

SI unit of electric displacement is C/m^2.

Superposition principle

According to this principle when more then one charges interact with unit positive charge, then the net intensity on the test charge is the vector sum of the intensity due to individual charge.

i.e. $$\vec{E} = \vec{E}_1 + \vec{E}_2 + ... + \vec{E}_n$$

Electric Potential Energy

Electrostatic Potential energy of a system of point charges is equal to the amount of work done to bring a text charge from infinity upto a certain point in the electric field of another charge.

Fig.

If two charges are assumed at a distance r_{12}.

Then electrostatic potential energy will be equal to the work done to bring charge q_2 from infinity to point B against the electric field of q_1 as

$$W = \int_{\infty}^{r_{12}} \vec{F}.\,dx$$

$$= -\int_{\infty}^{r_{12}} q_2 \vec{E}.\,dx$$

$$= -q_2 \int_{\infty}^{r_{12}} \frac{q_1}{4\pi\,\varepsilon_0\,x^2}\,dx$$

$$= \frac{1}{4\pi\varepsilon_0}\,q_1 q_2 \left[\frac{1}{x}\right]_{\infty}^{r_{12}}$$

$$W = \frac{1}{4\pi\varepsilon_0}\,\frac{q_1 q_2}{r_{12}} \qquad \qquad ...(1)$$

This work done is stored in the system as electrostatic potential energy U.

So
$$U = \frac{1}{4\pi\varepsilon_0}\,\frac{q_1 q_2}{r_{12}} \qquad \qquad ...(2)$$

if there is a system formed by more than two charges then the potential energy of the system is

$$U = \frac{1}{4\pi\varepsilon_0}\left[\frac{q_1 q_2}{r_{12}} + \frac{q_1 q_3}{r_{13}} + \frac{q_2 q_3}{r_{23}} ....\right] \qquad \qquad ...(3)$$

Electric Potential

It is the amount of work done to bring a unit positive charge from infinity to any point in the electric field of another charge.

i.e. It is the electrostatic potential energy per unit test charge

i.e.
$$V = \frac{W}{q_2}$$

$$= \frac{1}{4\pi\varepsilon_0} \frac{q_1}{r_{12}}$$

The amount of work done to bring a unit positive test charge from point B to point A against the electrostatic field of charge q_1 is known as electric potential difference.

i.e.
$$\int_B^A dV = \int_B^A \vec{E}.\overrightarrow{dr}$$

so
$$V_A - V_B = \frac{q_1}{4\pi\varepsilon_0}\left[\frac{1}{r_A} - \frac{1}{r_B}\right]$$

Electric Flux

The flux of any vector $\left(\vec{A}\right)$ associated with any elementary area $\left(\overrightarrow{ds}\right)$ measures the flow of the vector $\left(\vec{A}\right)$ through the elementary area vector is normal direction.

Mathematically

Flux
$$\phi = \vec{A}.\overrightarrow{ds}$$

$$= \vec{A}.\hat{n}\, ds$$

where $\hat{n}$ is the unit vector normal to ds.

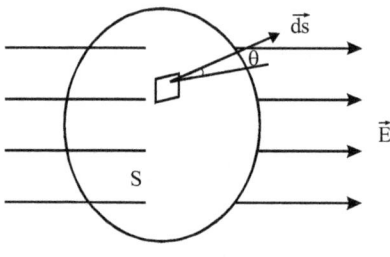

Fig.

When electric field lines moves in a certain direction through any area, then total number of electric lines of force passing normally through the small surface is called electric flux.

So electric flux

$$d\phi = \vec{E}.\overrightarrow{ds}$$

total flux through the entire surface

$$\phi = \oint_s \vec{E} \cdot \overrightarrow{ds} = \oint_s E \, ds \cos \theta$$

where θ is angle between $\vec{E}$ and $\overrightarrow{ds}$.

If the vector is directed outward form the surface, flux is +ve otherwise it is negative.

Conservative force field

When the work done on a particle is independent of the path so that total work done around the closed path is zero, the responsible force is called as conservative force and field developed is called the conservative field.

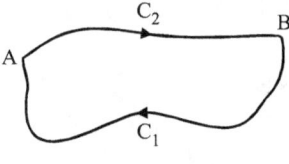

Fig.

Let us assume a particle moving from A to B via C_2 and returns back through C_1 in a conservative field of force $\vec{F}$.

total work done around the closed path AC_2BC_1A is zero

$$\oint_{AC_2BC_1A} \vec{F} \cdot \overrightarrow{dr} = \oint_{AC_2B} \vec{F} \cdot \overrightarrow{dr} + \oint_{BC_1A} \vec{F} \cdot \overrightarrow{dr} = 0$$

so

$$\oint_{AC_2B} \vec{F} \cdot \overrightarrow{dr} = \oint_{AC_1B} \vec{F} \cdot \overrightarrow{dr} = 0$$

so work done is independent of path.

So for conservative for $\oint_C \vec{F} \cdot \overrightarrow{dr} = 0$.

using stoke's theorem

$$\oint_C \vec{F} \cdot \overrightarrow{dr} = \oint_s (\vec{\nabla} \times \vec{F}) \cdot \overrightarrow{ds}$$

So for conservative force

$$\oint_s (\vec{\nabla} \times \vec{F}) \cdot \overrightarrow{ds} = 0$$

or $$\vec{\nabla} \times \vec{F} = 0 \qquad \qquad ...(1)$$

This is the condition for conservative force field.

for electric potential V (scalar potential)

curl grad V = 0

$$\boxed{\vec{\nabla} \times \vec{\nabla} V = 0} \qquad \qquad ...(2)$$

We can compare equation (1) & (2) and can conclude that force $\vec{F}$ should be a gradient of scalar potential V which is negative as per experimental evidence.

so for conservative field

$$\boxed{\vec{F} = -\vec{\nabla} V} \qquad \qquad ...(3)$$

Curl of electric field $\left(\vec{\nabla} \times \vec{E} \right)$

as electric field $$\vec{E} = \frac{1}{4\pi\varepsilon_0} \frac{q}{r^2} \hat{r} \qquad \qquad ...(i)$$

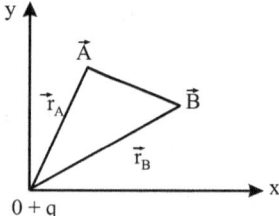

Fig.

line integral of electric field i.e. potential difference between two points A and B will be

$$\int_{A}^{B} \vec{E} . \overrightarrow{dr} = \frac{1}{4\pi\varepsilon_0} \int_{r_A}^{r_B} \frac{q}{r^2} dr$$

or $$\int_{r_A}^{r_B} \vec{E} . \overrightarrow{dr} = \frac{q}{4\pi\varepsilon_0} \left(\frac{1}{r_A} - \frac{1}{r_B} \right) \qquad \qquad ...(ii)$$

if $\overrightarrow{r_A} = \overrightarrow{r_B}$ i.e. A and B are same point.

then $$\oint_C \vec{E} \cdot \overrightarrow{dr} = 0$$

from stoke's law using $$\oint_C \vec{E} \cdot \overrightarrow{dr} = \int_s (\vec{\nabla} \times \vec{E}) \cdot \overrightarrow{ds}$$

$$\int_s (\vec{\nabla} \times \vec{E}) \cdot \overrightarrow{ds} = 0$$

or $$\vec{\nabla} \times \vec{E} = 0$$

Hence electric field is conservative.

Gauss' law of electrostatics

Gauss law states that the total electric flux (ϕ_E) through any closed surface is equal to $\dfrac{1}{\varepsilon_0}$ times of the total charge enclosed by the surface.

If any arbitrary shaped closed surface S encloses a charge q, then the electric flux through the closed surface is

$$\phi_E = \oint_s \vec{E} \cdot \overrightarrow{ds} = \frac{q}{\varepsilon_0}$$

Gauss law gives a relation between electric flux and charge.

The imaginary surface, where the magnitude of electric field intensity must be same and normal to the surface, is called Gaussian surface. for any medium

$$\phi_E = \oint_s \vec{E} \cdot \overrightarrow{ds} = \frac{q}{\varepsilon}$$

where ε is the permittivity of the medium. as

$$\vec{\Delta} = \varepsilon \vec{E}$$

$$\phi_E = \oint_s \vec{D} \cdot \overrightarrow{ds} = q$$

Proof :

Let us consider a point charge +q situated at 0 enclosed by the surface s.

if $\vec{E}$ is the electric field intensity at P due to charge +q at O along the direction OP

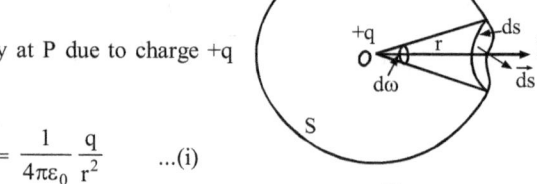

Fig.

$$E = \frac{1}{4\pi\varepsilon_0}\frac{q}{r^2} \quad ...(i)$$

then outward flux through S

$$\phi_E = \oint_s \vec{E}.\overrightarrow{ds} = \oint_s E\, ds\cos\theta = \frac{q}{4\pi\varepsilon_0}\oint\frac{ds\cos\theta}{r^2} \quad ...(ii)$$

here $\oint\dfrac{ds\cos\theta}{r^2} = \int dw$ is the solid angle subtended by the entire closed surfaces S at 0 and its value is 4π.

so $$\phi_E = \oint_s \vec{E}.\overrightarrow{ds} = \frac{q}{4\pi\varepsilon_0}\times 4\pi$$

$$\oint_s \vec{E}.\overrightarrow{ds} = \frac{q}{\varepsilon_0} \text{ which proves the Gauss' theorem.}$$

Gauss theorem can also be stated as the surface integral of the electric field over a closed surface (S) is equal to $\dfrac{1}{\varepsilon_0}$ times of the net charge (q) enclosed by the surfaces.

Differential form of Gauss' law

If charge q is distributed over a volume V enclosed by a surface S, then from the definition of volumetric charge distribution, the total charge enclosed by the volume is

$$q = \int_V \rho\, dV \quad\quad ...(i)$$

and Guass' law of electrostatics states that

$$\oint_s \vec{E}.\overrightarrow{ds} = \frac{q}{\varepsilon_0} \quad\quad ...(ii)$$

but from Gauss' divergence theorem

$$\oint_s \vec{E}.\vec{ds} = \int_V (\vec{\nabla}.\vec{E}) \, dV \qquad ...(iii)$$

So using (iii) & (i) in (ii)

$$\int_V (\vec{\nabla}.\vec{E}) \, dV = \frac{q}{\varepsilon_0} = \frac{1}{\varepsilon_0} \int_V \rho dV \qquad ...(iv)$$

so

$$\boxed{\vec{\nabla}.\vec{E} = \frac{\rho}{\varepsilon_0}} \qquad ...(v)$$

equation (v) is the differential form of Gauss' law of electrostatics. According to this the divergence of electric field at any point is $\dfrac{1}{\varepsilon_0}$ times the charge density at the point.

as

$$\vec{D} = \varepsilon_0 \vec{E}$$

then

$$\vec{\nabla}.(\varepsilon_0 \vec{E}) = \rho$$

so

$$\boxed{\vec{\nabla}.\vec{D} = \rho} \qquad ...(vi)$$

This is the differential form of Gauss' law in electrostatics in dielectric medium.

Limitation

(a) Here direction of electric field can not be measured as electric flux is a scalar quantity.

(b) This can only be used to find flux through regular/symmetric shaped bodies.

Coulomb's law from Gauss law

When two charges q_1 and q_2 are placed near to each other at some distance r then a gaussian surface of radius r can be drawn with q_1 at its centre A.

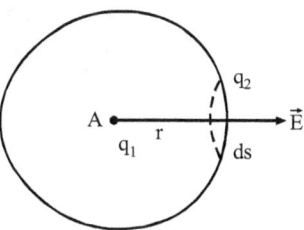

then from Gauss theorem

$$\oint_s \vec{E}.\vec{ds} = \frac{q_1}{\varepsilon_0} \qquad ...(i)$$

or $$\oint_s E\, ds \cos 0 = \frac{q_1}{\varepsilon_0}$$

$$E\oint ds = \frac{q_1}{\varepsilon_0} \Rightarrow E.4\pi r^2 = \frac{q_1}{\varepsilon_0}$$

$$\Rightarrow \qquad E = \frac{q_1}{4\pi\varepsilon_0 r^2} \qquad \qquad ...(ii)$$

and the force acting on q_2 due to the field

$$F = q_2\, E = \frac{q_1 q_2}{4\pi\varepsilon_0 r^2} \qquad \qquad ...(iii)$$

which is mathematical form of coulomb's law.

Some applications of Gauss' Law of Electrostatics

According to Gauss law $\oint_s \vec{D}.\vec{ds} = Q$

$\vec{D}.\vec{ds} = Dds$ when D is perpendicular to the surface.

$\vec{D}.\vec{ds} = 0$ when D is tangential to the surface.

(a)   Field due to a point charge

Suppose a point charge is placed at origin then a sphere with centre at origin can be considered as Gaussian surface.

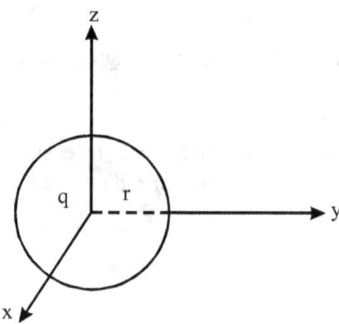

Fig.

so $\qquad \int d\phi$ = total flux outward

$$= \oint_s \vec{D}.\overrightarrow{ds} = q$$

$$\Rightarrow \qquad \phi = \oint_s \varepsilon_0 \vec{E}.\overrightarrow{ds} = q$$

$$\Rightarrow \qquad \varepsilon_0 \oint E \; ds = q$$

$$\Rightarrow \qquad \varepsilon_0 \, 4\pi r^2 . E = q$$

$$\Rightarrow \qquad E = \frac{q}{4\pi\varepsilon_0 r^2}$$

and $\qquad \boxed{\vec{E} = \frac{q}{4\pi \; \varepsilon_0 r^2} \; \hat{r}}$ $\qquad$ ...(i)

(b) Field due to infinite line charge

The Gaussian surface for line charge can be a cylindrical surface with $\vec{E}$ radially outward over its surface.

The total outward flux is

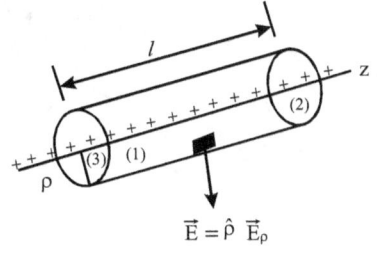

$$E = \hat{\rho} \; \vec{E}_\rho$$
$$\overrightarrow{dS} = \hat{\rho} \; dS$$

$$\underset{(1)}{\oint \vec{D}.\overrightarrow{ds}} + \underset{(2)}{\oint \vec{D}.\overrightarrow{ds}} + \underset{(3)}{\oint \vec{D}.\overrightarrow{ds}} = \lambda l = q$$

where λ is linear charge density.

But $\qquad \underset{(2)}{\oint \vec{D}.\overrightarrow{ds}} = \underset{(3)}{\oint \vec{D}.\overrightarrow{ds}} = 0$ as $\theta = 90°$ between $\vec{D}$ & $\overrightarrow{ds}$.

$$\Rightarrow \qquad \underset{1}{\oint \vec{D}.\overrightarrow{ds}} = \lambda l$$

$$\Rightarrow \qquad \left(\varepsilon_0 \vec{E}_\rho \; \hat{\rho}\right).\hat{\rho} \; \overrightarrow{ds} = \lambda l$$

$$\Rightarrow \qquad \varepsilon_0 \, \vec{E}_\rho \, 2\pi\rho l = \lambda l$$

$$\Rightarrow \qquad \boxed{\vec{E}_\rho = \frac{\lambda}{2\pi \, \varepsilon_0 \, \rho} \, \hat{\rho}} \qquad \qquad ...(ii)$$

(c) Field of a charged cloud

Consider a spherical cloud of radius r_0 with uniform charge density ρ C/m^3.

(i)   Field outside the cloud i.e. $r > r_0$ –

for S_1

$$\oint_{S_1} \vec{D}.\overrightarrow{ds} = q$$

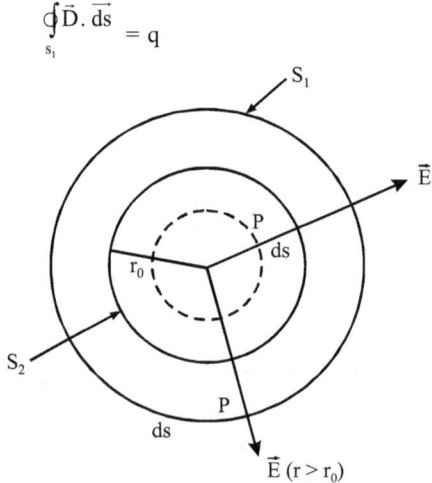

Fig.

or $$\varepsilon_0 \oint_{S_1} \vec{E}.\overrightarrow{ds} = \rho \int dv$$

$$= \rho \left(\frac{4}{3}\pi r_0^3 \right)$$

or $$\varepsilon_0 E \, 4\pi r^2 = \rho \, \frac{4}{3}\pi r_0^3$$

$$E = \frac{\rho \left(4/3 \, \pi r^3 \right)}{4\pi \, \varepsilon_0 r^2} = \frac{\rho r_0^3}{3\varepsilon_0 r^2} \qquad \qquad ...(2)$$

or $$\vec{E} = E\,\hat{n}$$

$$\vec{E} = \frac{\rho\,r_0^3}{3\varepsilon_0 r^2}\,\hat{n}$$

(ii) Field inside the cloud i.e. $r < r_0$ –

for closed surface s_2

$$\oint_{s_2} \vec{D}.\overrightarrow{ds} = \rho \int_{V_2} dV$$

$$\varepsilon_0 \oint_{s_2} \vec{E}.\overrightarrow{ds} = \rho \left[\frac{4}{3}\pi r^3\right]$$

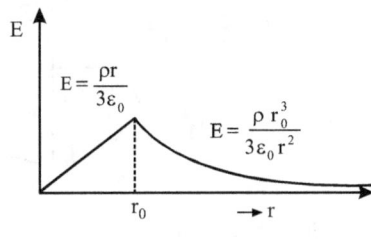

$$\varepsilon_0\, E\, 4\pi r^2 = \rho\, \frac{4}{3}\,\pi r^3$$

$$E = \frac{\rho r}{3\varepsilon_0}$$

Fig.

Hence the variation of electric field is as below

(d) Infinite sheet of charge

Suppose there is an infinite sheet of uniform charge σ C/m^2.

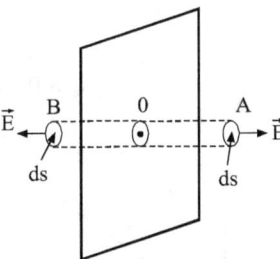

Fig.

the total flux ϕ remains over the two ends of Gaussian cylindrical surface

$$= 2\vec{E}.\overrightarrow{ds}$$

$$= 2\, E\, ds$$

So $\oint_s 2E\ ds = \dfrac{q}{\varepsilon_0} = \dfrac{\sigma\ ds}{\varepsilon_0} = 2E\ ds$

or $$\boxed{E = \dfrac{\sigma}{2\varepsilon_0}}$$

(e)   Field due to a parallel plate capacitor–

A parallel plate capacitor is a combination of two parallel plates separated by some distance.

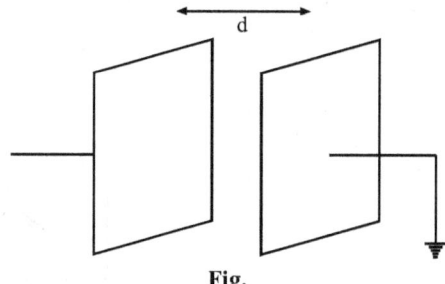

Fig.

So $E = 2E_1$ where E_1 is electric field due to single plate.

$$= 2.\dfrac{\sigma}{2\varepsilon_0}$$

$\Rightarrow$ $$\boxed{\vec{E} = \dfrac{\sigma}{\varepsilon_0}\hat{n}}$$

Potential Gradient

Electric field $\vec{E}$ due to +q charge placed at origin 0.

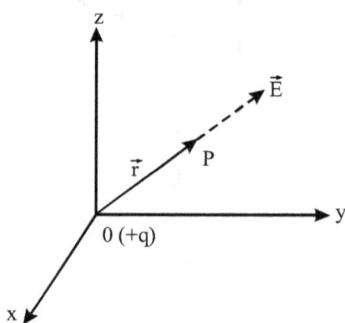

$$\vec{E}(r) = \frac{q}{4\pi\,\varepsilon_0 r^2}\,\hat{r} = \frac{q\,\hat{r}}{4\pi\,\varepsilon_0 r^2} \qquad ...(i)$$

if $\hat{r}$ is the position vector of point (x, y, z)

then
$$r = \sqrt{x^2 + y^2 + z^2}$$

and
$$\vec{\nabla}\left(\frac{1}{r}\right) = \vec{\nabla}\left(\frac{1}{\sqrt{x^2 + y^2 + z^2}}\right)$$

$$= \vec{\nabla}\left(\left(x^2 + y^2 + z^2\right)^{-1/2}\right)$$

$$= \frac{-\left(x\hat{i} + y\hat{j} + z\hat{k}\right)}{\left(x^2 + y^2 + z^2\right)^{3/2}}$$

$$= -\frac{\vec{r}}{r^3} \qquad ...(ii)$$

so using (ii) in (i)

$$\vec{E}(r) = -\frac{q}{4\pi\varepsilon_0}\,\vec{\nabla}\left(\frac{1}{r}\right)$$

$$= -\vec{\nabla}\left(\frac{q}{4\pi\varepsilon_0 r}\right)$$

$$= -\vec{\nabla}V$$

hence
$$\boxed{\vec{E}(r) = -\vec{\nabla}V}$$

(a) $-$ve sign indicates that the direction of $\vec{E}$ is along the direction of decreasing V.

(b) When $\vec{E} = 0$ then $\vec{\nabla}V = 0 \Rightarrow V =$ constant. so electric potential will have constant value not compulsorily zero.

Poisson's and Laplace Equation

According to the differential form of Gauss' law

$$\bar{\nabla}.\vec{E} = \frac{\rho}{\varepsilon_0} \qquad \text{...(i)}$$

and

$$\vec{E} = -\vec{\nabla}V \qquad \text{...(ii)}$$

using (ii) in (i)

$$\vec{\nabla}.\left(-\vec{\nabla}V\right) = \frac{\rho}{\varepsilon_0}$$

or

$$\boxed{\nabla^2 V = -\frac{\rho}{\varepsilon_0}} \qquad \text{...(iii)}$$

or

$$\text{div grad } V = -\frac{\rho}{\varepsilon_0}$$

This equation (iii) is known as Poisson equation. Any static electric field should satisfy this equation which gives a relation between potential and charge density at any point in an electric field in space.

For a charge free region $\rho = 0$

$$\boxed{\nabla^2 V = 0} \qquad \text{...(iv)}$$

equation (iv) is known as Laplace's equation.

The solution of this equation gives the potential in charge free space.

as $\nabla^2 V = 0 \Rightarrow V = $ constant, so potential remains constants at all points.

The Laplace equation in three coordinates are

(i)   Cartessian coordinates

$$\nabla^2 V = \frac{\partial^2 V}{\partial x^2} + \frac{\partial^2 V}{\partial y^2} + \frac{\partial^2 V}{\partial z^2} = 0 \qquad \text{...(v)}$$

(ii)   Cylindrical coordinates

$$\nabla^2 V = \frac{1}{\rho}\frac{\partial}{\partial \rho}\left(\rho\frac{\partial V}{\partial \rho}\right) + \frac{1}{\rho^2}\frac{\partial^2 V}{\partial \phi^2} + \frac{\partial^2 V}{\partial z^2} = 0 \qquad \text{...(vi)}$$

(iii)   Spherical coordinates

$$\nabla^2 V = \frac{1}{r^2}\frac{\partial}{\partial r}\left(r^2\frac{\partial V}{\partial r}\right) + \frac{1}{r^2 \sin\theta}\frac{\partial}{\partial\theta}\left(\sin\theta\frac{\partial V}{\partial\theta}\right)$$

$$+\frac{1}{r^2 \sin^2\theta}\frac{\partial^2 V}{\partial\phi^2} = 0 \qquad\qquad ...(viii)$$

Applications of Laplace equation in Electrostatics–

(i)   A parallel plate capacitor–

Consider a parallel plate capacitor having its plates at z = 0 and z = d with upper plates potential at V_1 and lower plate grounded as shown

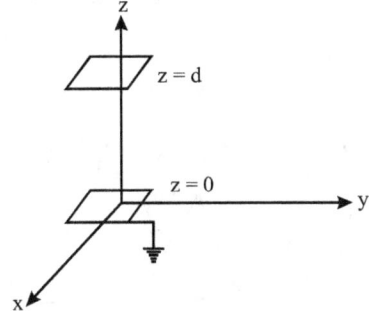

Fig.

here X – Y component of potential are zero so

$$\frac{\partial^2 V}{\partial x^2} = \frac{\partial^2 V}{\partial y^2} = 0$$

hence from Laplace equation in Cartesian coordinates

$$\frac{\partial^2 V}{\partial z^2} = 0$$

Integrating $\dfrac{\partial V}{\partial z} = C$ and again

$$V = Cz + D \qquad\qquad ...(i)$$

Using boundary conditions

at $z = 0$, $V = 0$,   so $D = 0$

and at $z = d$, $V = V_1$ so $C = \dfrac{V_1}{d}$ ...(ii)

so using (ii) in (i)

$$\boxed{V = \dfrac{V_1 z}{d}}$$ ...(iii)

(ii)   A spherical capacitor

Consider any point P inside a sphere at a distance r from the centre of the sphere. The variation of potential V is considered only along radial direction.

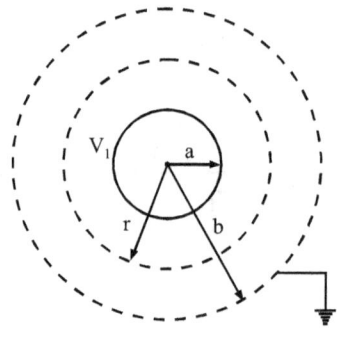

Fig.

So Laplace equation in spherical coordinates as

$$\nabla^2 \equiv \frac{1}{r^2}\frac{\partial}{\partial r}\left(r^2 \frac{\partial V}{\partial r}\right) = 0$$

or $\dfrac{\partial}{\partial r}\left(r^2 \dfrac{\partial V}{\partial r}\right) = 0$ ...(i)

Integrating

$$r^2 \frac{\partial V}{\partial r} = \text{constants} = C$$

$$\Rightarrow \qquad \frac{\partial V}{\partial r} = \frac{C}{r^2}$$

Integrating

$$V = -\frac{C}{r} + D \qquad \qquad ...(ii)$$

applying boundary conditions

at r = b, V = 0

$$0 = -\frac{C}{b} + D$$

or

$$D = \frac{C}{b}$$

so

$$V = -\frac{C}{r} + \frac{C}{b} = C\left[\frac{1}{b} - \frac{1}{r}\right] \qquad \qquad ...(iii)$$

at r = a, V = V_1

so

$$V_1 = C\left[\frac{1}{b} - \frac{1}{a}\right]$$

or

$$C = \frac{V_1}{\left(\frac{1}{b} - \frac{1}{a}\right)} \qquad \qquad ...(iv)$$

Using (iii) & (iv) in (ii)

$$V = \frac{V_1}{\left(\frac{1}{b} - \frac{1}{a}\right)}\left(\frac{1}{b} - \frac{1}{r}\right)$$

or

$$V = \frac{V_1}{k}\left(\frac{1}{b} - \frac{1}{r}\right) \quad \text{where} \quad k = \frac{1}{b} - \frac{1}{a} = \frac{a - b}{ab} \qquad ...(v)$$

Which is the expression of potential inside a spherical capacitor.

(iii)   Cylindrical capacitor

Consider any point P inside a cylinder at distance r with variation of potential only along the radial direction only. So Laplace equation in cylindrical coordinates reduces to

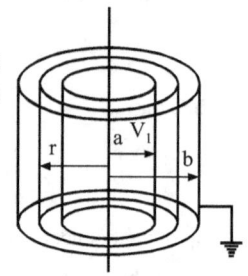

$$\nabla^2 V \equiv \frac{1}{r}\frac{\partial}{\partial r}\left(r\frac{\partial V}{\partial r}\right) = 0$$

or
$$\frac{\partial}{\partial r}\left(r\frac{\partial V}{\partial r}\right) = 0 \qquad ...(i)$$

Integrating (i) $\Rightarrow$
$$r\frac{\partial V}{\partial r} = \text{constant} = C$$

or
$$\frac{\partial V}{\partial r} = \frac{C}{r}$$

Integrating again
$$V = C \ln r + D \qquad\qquad ...(ii)$$

applying boundary conditions

at $r = b,\ V = 0$

$$0 = C \ln r + D$$

or
$$D = -C \ln b \qquad\qquad ...(iii)$$

Using (iii) in (ii)
$$V = C \ln r - C \ln b$$

$$= C \ln \frac{r}{b} \qquad\qquad ...(iv)$$

when $r = a,\ V = V_1$

$$V_1 = C \ln \frac{a}{b}$$

or
$$C = \frac{V_1}{\ln \dfrac{a}{b}} \qquad\qquad ...(v)$$

So from (ii)

$$V = \frac{V_1 \ln \dfrac{r}{b}}{\ln \dfrac{a}{b}} \qquad \qquad ...(vi)$$

This gives the potential at a point inside the cylindrical capacitor with a $< r <$ b.

Electric current

There are large number of free electrons available due to overlapping of conduction band and valence band in metals. For metals energy band gap $\Delta E_g = 0$ so free electrons move inside the metal in all directions and the net rate of movement is zero. Hence in absence of any external electric field there is no net transfer of charge. but when an external electric field is applied then due to electrostatic force, these free electrons get accelerated and move towards opposite to the electric field. The rate of flow of electric charge through any cross section of the conductor is known as electric current (I).

i.e.
$$I = \frac{dq}{dt} \qquad \qquad ...(i)$$

If this current (I) is uniformly distributed across conductor of cross sectional area A then the current density at all points in this cross section is defined as the current flowing per unit area of cross section of the conductor.

$$J = \frac{I}{A} \quad \text{which is a vector quantity}$$

$$I = \oint_s \vec{J} . \vec{ds}$$

It represents the current flowing through the entire surface s.

Equation of Continuity

If ρ is the charge density, then in a small volume element dV, then total charge in this volume

$$q = \int_V \rho \, dV \qquad \qquad ....(i)$$

Current I is the rate of decrease of charge from the volume of the conductor

$$I = -\frac{dq}{dt} \qquad \qquad ...(ii)$$

so
$$I = -\frac{d}{dt} \int_V \rho \, dV = -\int_V \frac{\partial \rho}{\partial t} dV \qquad \text{...(iii)}$$

If $\bar{j}$ is the current density then current is

$$I = \oint_s \vec{J} \cdot \overrightarrow{ds} \qquad \text{...(iv)}$$

so from (iii) & (iv)

$$\oint_s \vec{J} \cdot \overrightarrow{ds} = -\int_V \frac{\partial \rho}{\partial t} dV \qquad \text{...(v)}$$

Form Gauss's Divergence theorem

$$\oint_s \vec{J} \cdot \overrightarrow{ds} = \int_V \left(\vec{\nabla}.\vec{J}\right) dV \qquad \text{...(vi)}$$

So from (v) using (vi)

$$\int_V \left(\vec{\nabla}.\vec{J}\right) dV = -\int_V \frac{\partial \rho}{\partial t} dV$$

or
$$\int_V \left(\vec{\nabla}.\vec{J} + \frac{\partial \rho}{\partial t}\right) dV = 0 \qquad \text{...(vii)}$$

or
$$\boxed{\vec{\nabla}.\vec{J} + \frac{\partial \rho}{\partial t} = 0} \qquad \text{...(viii)}$$

This represents the equation of continuity which provides a relation between charge density and current density.

Equation (viii) implies that decrease of charge inside a volume of a conductor with time is equal to the flow of charge from that surface enclosing the volume, which satisfies the conservation of charge.

Drift Velocity

Let us consider a conductor of length l and cross section area A.

In a conductor the electrons are in random motion in all directions due to which the net effect is zero. But when an electric field is applied across the conductor free

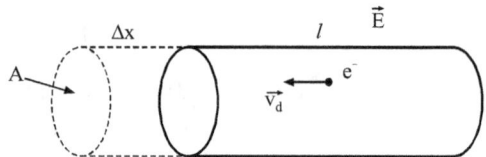

electrons are drifted towards the electric field with a certain velocity. During this motion electron collides with other particle and again comes to rest and then under electrostatic force again starts to move in any random direction. The time elapsed between two successive collisions is known as relaxation time. Overall electron drifts towards a certain direction of electric field.

The drift velocity is given by

$$v_d = (u_1 + u_2 + ...) + \frac{a(\tau_1 + \tau_2 + \tau_3 + ...)}{n}$$

where u_1, u_2, u_3 .... are initial velocities of electron after collision, a is the acceleration generated due to electrostatic force and τ_1, τ_2, τ_3 are relaxation times during two successive collisions.

$$u_1 = u_2 = u_3 \, ... = 0$$

so $$v_d = \frac{a(\tau_1 + \tau_2 + \tau_3 + ...)}{n}$$

$$= a\tau \text{ where } \tau \text{ is the mean relaxation time.} \qquad ...(i)$$

the electrostatic force

$$\vec{F} = e\vec{E} \qquad ...(ii)$$

so $$eE = a\tau \Rightarrow a = \frac{eE}{m} \qquad ...(iii)$$

using (iii) in (i)

$$v_d = \frac{eE}{m}\tau \qquad ...(iv)$$

due to this drifting of electrons a current is setup across the conductor known as steady current if there are n free electrons per unit volume then the total no. of free electrons passing through cross section area A in time Δt.

$$N = n\,A\,x$$

$$= n\,A\,v_d\,\Delta t \qquad ...(v)$$

total charge passing through

$$\Delta q = ne\,A\,v_d\,\Delta t \qquad ...(vi)$$

hence the rate of flow of charge i.e. steady current

$$J = \frac{\Delta q}{\Delta t}$$

$$= \frac{ne \, A \, v_d \, \Delta t}{\Delta t}$$

$$I = ne \, A \, v_d \qquad\qquad ...(vii)$$

using $\qquad\qquad v_d = \dfrac{eE}{m} \tau$ in (vii)

$$I = \frac{ne^2 AE}{m} \tau \qquad\qquad ...(viii)$$

According to Ohm's claw the potential difference V and resistance R is related with I as

$$V = IR$$

and potential difference $\qquad V = El$

$$\Rightarrow \qquad\qquad E = \frac{V}{l} \qquad\qquad ...(ix)$$

so $\qquad\qquad I = \dfrac{ne^2 \, A \, V}{m \, l} \tau$

or $\qquad\qquad \dfrac{V}{I} = \dfrac{m \, l}{n \, e^2 \, \tau \, A} = R$

so $\qquad\qquad R = \dfrac{m}{ne^2 \tau} \dfrac{l}{A} = \rho \dfrac{l}{A} \qquad\qquad ...(x)$

where $\qquad\qquad \rho = \dfrac{m}{ne^2 \tau}$ is the resistivity of the conductor.

$$\text{Current density } \vec{j} = \frac{I}{A} = \frac{ne^2 \, \vec{E}}{m} \tau$$

or $\qquad\qquad \boxed{\vec{J} = \sigma \vec{E}} \qquad\qquad ...(xi)$

where $\qquad\qquad \sigma = \dfrac{ne^2 \vec{E}}{m}$ is conductivity of the conductor.

Average drift velocity per unit electric field is called as mobility

as $\qquad\qquad v_d \propto E$

so $\qquad\qquad v_d = \mu_e E$

or
$$\mu_e = \frac{v_d}{E}$$

Where μ_e is the mobility of electron.

DIELECTRICS

As we know that conductors have a very large number of free electrons so that electrons can move in the material without any problem of electrostatic force as it is negligible and there is no requirement to apply external electric field to move these free electrons.

On the other hand insulators do not have any free electrons and conduction is not possible even after the application of external electric field, but there are certain materials which are insulators in all conditions but when some external field is applied on them they get polarized and show conducting properties and an electric field is setup inside the material so as to oppose the external electric field. These are known as dielectric materials. So, dielectric materials are those which are electrical insulators and an electrical field can be setup in that without much dissipation of power.

This property of dielectric materials is used to construct radio frequency transmission lines and to develop capacitors especially at radio frequencies. Insulating materials used to resist the current flow, while dielectric materials are used to store electrical energy.

Hence dielectric material in defined as a special material which remains insulator under almost all conditions and can be polarized and electrostatic field can be setup in it for a long time.

e.g. parafinwax, air, paper, glass, mica etc.

The resistivity of the dielectric material is in the range of 10^6 Ω-m to 10^{16} Ω-m.

Effect of a dielectric on the behaviour of a capacitor

Let us consider a parallel plate capacitor with vacuum between the plates with capacitance C_0 given by

$$C_0 = \frac{Q_0}{V} = \frac{\varepsilon_0 A}{d}$$

where A is the plate area and ε_0 is the electrical permittivity of vacuum.

Whenever a dielectric slab is inserted in between the plates, keeping V same then charge increases from Q_0 to Q on the plates.

so capacitance increases by the ratio Q to Q_0.

then
$$\varepsilon_r = \frac{Q}{Q_0} = \frac{C}{C_0}$$

where ε_r is the relative permittivity.

$$\varepsilon_r = \frac{\varepsilon_m}{\varepsilon_0}$$ ratio of permittivity in the medium to free space.

$$\varepsilon_0 = \frac{1}{c^2 \mu_0} = 8.854 \times 10^{-12} \text{ F/m}$$

where C is speed of light and μ_0 is permeability of free space.

This relative permittivity is defined as the dielectric constant i.e. $k = \frac{\varepsilon}{\varepsilon_0}$.

Hence the dielectric constant of a material is the ratio of the capacitance of a given capacitor filled completely with that material to the capacitance in vacuum or in other words, the ratio of permittivity of medium to that of the vacuum is known as dielectric constant.

It is found to be independent of the shape and dimension of the capacitor.

Dielectric constants of some materials (Table)

S. No.	Material	Dielectric constant k	Temp (F)
1.	Vacuum	1	-
2.	Air dry	1	68
3.	Polyethylene	2.25	68
4.	Paper	3	68
5.	Teflon	2	75
6.	Paraffin	2.3	75
7.	Petroleum oil	2	75
8.	Water	80	68
9.	Mica	7	75
10.	Glycerin	47	68
11.	Rubber	7	68
12.	Methyl Alcohol	30	68
13.	Barium titanate	1200	68

Dipole moment and polarization

An electrical dipole is an arrangement where two equal and opposite charges are held at very small distance to each other.

When any dielectric substances is kept inside any electric field then the effect of this field induces electrical dipoles in the material and try to align them into field direction. In addition to the generation of new electric dipoles, external field also try to align already existed dipoles and we have the combined effect. This total effect of an external electric field on any dielectric material is called polarization of the dielectric substance.

Any electrical dipole is characterized by its dipole moment which is the product of the magnitude of the charge and the separation between the centre of masses of +ve and −ve charges

hence dipole moment.

$$\vec{p} = q\,\vec{r} \qquad\qquad ...(1)$$

It is directed from negative to positive charge and hence it is a vector quantity. Its unit is debye (D).

where,   1 debye = 10^{-18} stat C. cm.

$$\approx 3.33 \times 10^{-30} \text{ C.m}$$

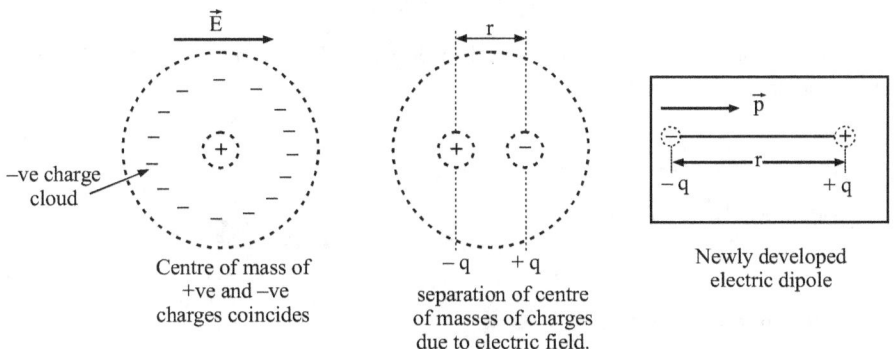

Centre of mass of
+ve and −ve
charges coincides

separation of centre
of masses of charges
due to electric field.

Newly developed
electric dipole

The electric dipole moment per unit volume is called polarization or polarization density $\left(\vec{P}\right)$. It is always directed from negative charge to positive charge.

If there are N atoms per unit volume than

$$\vec{p} = N\vec{p} \qquad\qquad ...(ii)$$

where $\vec{p}$ is the electric dipole moment of individual atom.

The dielectric molecules become polarized when it is placed under some external electric field. The polarization vector $(\vec{p})$ is proportional to electric field experienced by the dielectric molecules so

$$\vec{P} \propto \vec{E}$$

$$\vec{P} = \chi \, \varepsilon_0 \, \vec{E} \qquad\qquad\qquad ...(iii)$$

Where χ is called the electrical susceptibility of the dielectric material. It is equal to the ratio of polarization per unit volume to electrical intensity in the dielectric.

Again, the net induced dipole moment $(\vec{p})$ of an atom of dielectric substance placed in an electric field is proportional to the applied field $(\vec{E})$ with its direction parallel to the field so that

$$\vec{p} \propto \vec{E} \quad \text{or} \quad \vec{p} = \alpha \vec{E} \qquad\qquad\qquad ...(iv)$$

where α is called atomic polarizability $\left(\alpha = \dfrac{\vec{p}}{\vec{E}}\right)$.

Hence atomic polarizability is equal to the induced dipole moment for an atom when electric field of unit strength is applied on it.

$$\alpha = \frac{p}{E} = \frac{C.m}{Vm^{-1}} = CV^{-1}m^2 = Fm^2$$

so form (ii)

$$\vec{P} = n \, \alpha \, \vec{E} \qquad\qquad\qquad ...(v)$$

As we know that polarization density

$$P = \frac{\text{Total dipole moment}}{\text{Volume of dielectric slab}}$$

$$= \frac{qd}{sd} = \frac{q}{s} = \sigma_p$$

so $\qquad\qquad\qquad P = \sigma_p \qquad\qquad\qquad ...(vi)$

When dielectric slab is placed inside a parallel plate capacitor, the effective electric field is reduced by

$$E = \frac{\sigma - \sigma_p}{\varepsilon_0} = \frac{\sigma}{\varepsilon_0} - \frac{\sigma_p}{\varepsilon_0}$$

$\Rightarrow$ $$E = E_0 - \frac{\sigma_p}{\varepsilon_0} \qquad \left(\text{or } E = E_0 - \frac{P}{\varepsilon_0} \right) \qquad ...(vii)$$

$\Rightarrow$ $$\varepsilon_0 E = \varepsilon_0 E_0 - \sigma_p$$

$\Rightarrow$ $$\varepsilon_0 E_0 = \varepsilon_0 E + \sigma_p$$

using (vi)

$$\varepsilon_0 E_0 = \varepsilon_0 E + P$$

But $\varepsilon_0 E_0$ is known as the electrical displacement vector D, so

$$\boxed{\vec{D} = \varepsilon_0 \vec{E} + \vec{P}} \qquad ...(viii)$$

As $$\chi = \frac{\vec{P}}{\varepsilon_0 \vec{E}}$$

$\Rightarrow$ $$\vec{P} = \chi \varepsilon_0 \vec{E}$$

placing in above (vii)

$$E = E_0 - \frac{\chi \varepsilon_0 E}{\varepsilon_0} = E_0 - \chi E$$

or $$\frac{E_0}{E} = 1 + \chi$$

$\Rightarrow$ $$\boxed{k = 1 + \chi} \qquad ...(ix)$$

where $k = \dfrac{E_0}{E}$ is the dielectric constant of the dielectric material.

TYPES OF DIELECTRICS

A molecule is assumed to be neutral where the algebraic sum of all the charges is zero. Any material can have a large number of molecules and each molecule consist of a nucleus and electrons. Depending on the separation between the centre of gravity of positive and negative charges the material can be classified as polar or non polar dielectric material.

(i)   Non-polar dielctric

A molecule in which the centre of gravity of positive and negative charges coincides, due to which the molecule does not possess any permanent dipole moment in the absence of external electric field. There are called non-polar molecule and the material as non-polar dielectric material. these generally have symmetrical arrangement.

Monoatomic molecules like He, Ne, Ar, Xe or molecules having two identical atoms like H_2, N_2, O_2, Cl_2 etc. are non-polar. CO_2 due to its linearly symmetric structure is also non-polar.

(ii)   Polar dielectrics

Here the centre of gravity of positive charge is finitely separated from that of negative charge, resulting in an electric dipole with some dipole moment. which is permanent.

Materials made up of these molecules are called polar dielectrics.

Polyatomic molecules like N_2O, H_2O, HCl, NH_3 etc. are polar.

Types of polarization

Some of the important polarizations are as under–

(i)   Electronic polarization

Here when the external field is applied, the electron clouds of atom are displaced with respect to the heavy nuclei within the dimensions of atom. This is called electronic polarization. It does not depend upon temperature.

$$\vec{P}_e = N\,\alpha_e\vec{E}$$

(ii)   Ionic polarization

It occurs only in some ionic crystals. In the presence of external electric field the positive and negative ions are displaced upto the point where ionic bonding force stop this displacement. Hence dipoles gets induced. There also do not depend upon temperature.

(iii) Orientational polarization

It applies only in polar dielectric materials. Generally, in absence of external electric field electric dipoles are so oriented randomly that their net effect becomes zero but in presence of electric field, these dipole try to rotate and align in the direction of electric field. This is known as orientation polarization which is dependent over temperature also.

The total polarization in the sum of all there effects as $\vec{P} = \vec{P}_e + \vec{P}_i + \vec{P}_0$

Relation between electronic polarizability and atomic radius

Let us consider an atom with atomic number Z and radius r.

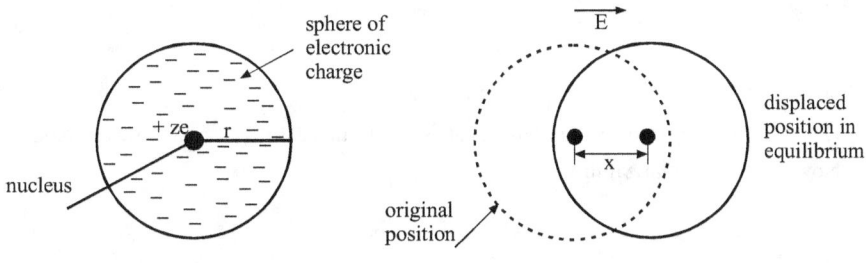

without any field

Electronic polarization under the effect of electric field $\vec{E}$

here charge density

$$\rho = \frac{-Ze}{\frac{4}{3}\pi r^3}$$

when this atom is placed inside any electric field $\vec{E}$, there is a displacement of x between nucleus and electron cloud. The responsible force

$$F = (Ze)\ E \qquad \qquad ...(i)$$

and the magnitude of coulomb's attractive force between nucleus and electron cloud is

$$F' = \frac{(ze) \times \text{charge enclosed in the sphere of radius x}}{4\pi\varepsilon_0 x^2}$$

$$= \frac{(ze)\left(\frac{4}{3}\pi x^3 \cdot \rho\right)}{4\pi\varepsilon_0 x^2}$$

$$= -\frac{z^2 e^2 x}{4\pi\varepsilon_0 r^3} \qquad \qquad ...(ii)$$

at equilibrium the nucleus is balanced, so total effect of force must be zero.

$$zeE = + \frac{z^2 e^2 x}{4\pi \, \varepsilon_0 \, r^3}$$

or
$$x = \frac{4\pi \, \varepsilon_0 \, r^3 E}{Ze} \qquad \qquad ...(iii)$$

which shows displacement $x \propto$ applied field E

so the induced electronic dipole moment

$$p = (ze)x$$

or
$$\vec{p} = 4\pi \, \varepsilon_0 \, r^3 \, \vec{E} \qquad \qquad ...(iv)$$

Hence induced electronic dipole moment is proportional to the applied field strength.

Now electronic polarizability

$$\alpha_e = \frac{\vec{p}}{\vec{E}}$$

$$\alpha_e = 4\pi \, \varepsilon_0 r^3 \qquad \qquad ...(v)$$

if there are n atoms per cubic meter, the total electronic polarization

$$\vec{P} = n\vec{p}$$

$$= n \alpha_e \vec{E}$$

$$\vec{P} = n\left(4\pi\varepsilon_0 r^3\right) \vec{E} \qquad \qquad ...(vi)$$

Gauss law in dielectric materials

Gauss law states that the total electric flux ϕ passing through a closed surface is equal to the total charge enclosed by that surface.

i.e.
$$\phi = \oint_s \vec{E}.\overrightarrow{ds} = \frac{q}{\varepsilon_0}$$

where q is the total charge enclosed

so
$$\oint_s \varepsilon_0 \vec{E}.\overrightarrow{ds} = q$$

or
$$\oint_s \vec{D}.\overrightarrow{ds} = q$$

SOLVED EXAMPLES

Q.1: Find the total charge contained within a sphere of radius 'r' and volume charge density proportional to radius.

Sol.: As volume charge density $\propto$ radius

so
$$\rho \propto r$$

$$\rho = Cr$$

C is constant of proportionality.

we know that differential volume element in spherical coordinate system

$$dV = r^2 \sin\theta dr\, d\theta\, d\phi$$

hence small charge element

$$d\theta = \rho \times dv$$

$$= Cr \times r^2 \sin\theta dr\, d\theta\, d\phi$$

$$= Cr^3 \sin\theta\, dr\, d\theta\, d\phi$$

hence total charge

$$Q = \int_0^r \int_{\theta=0}^{\pi} \int_{\phi=0}^{2\pi} C\, r^3 \sin\theta\, d\theta\, dr\, d\phi$$

$$= C \int_0^r r^3 dr \int_0^{\pi} \sin\theta\, d\theta \int_0^{2\pi} d\phi$$

$$= C \cdot \frac{r^4}{4} \cdot \cos\theta \Big|_0^{\pi}\; 2\pi$$

$$Q = \pi\, C r^4$$

Q.2 : Two unknown point charges are placed at a distance r apart. Find the ratio of the charges at a point on the line joining them where electric at a point on the line joining them where electric field strength becomes zero. Also comment about their natures.

Sol.: If q_1 and q_2 are the two charges.

then

Fig.

so
$$\frac{q_1}{4\pi\, \varepsilon_0 x^2} = \frac{q_2}{4\pi\, \varepsilon_0 (r-x)^2}$$

or
$$\frac{q_2}{q_1} = \left(\frac{r-x}{x}\right)^2$$

or
$$q_2 : q_1 = (r-x)^2 : x^2$$

as $\left(\dfrac{r-x}{x}\right)^2$ will always be +ve so both the charges will either +ve or –ve.

Q.3: Point charges 5nC and 2nC are located at (2, 0, 4) and (–3, 0, 5) respectively. Find

(a)   Force on 1 nC point charge located at (1, –3, 7).

(b)   Electric field at (1, –3, 7).

[RU 2006]

Sol.:

Let
$$q_1 = 5 \text{ nC at } (2, 0, 4)$$

so
$$\vec{r_1} = 2\hat{i} + 4\hat{k}$$

and
$$q_2 = 2\text{nC at} (-3, 0, 5)$$

so
$$\vec{r_2} = -3\hat{i} + 5\hat{k}$$

and
$$q = 1 \text{ nC at } (1, -3, 7)$$

so
$$\vec{r} = \hat{i} - 3\hat{j} + 7\hat{k}$$

so that
$$\vec{r} - \vec{r_1} = \hat{i} - 3\hat{j} + 3\hat{k}$$

and
$$\vec{r} - \vec{r}_2 = 4\hat{i} - 3\hat{j} + 2\hat{k}$$

(i)   According to superposition principle

$$\vec{F} = \sum_{i=1}^{2} k \frac{qQ_i}{|r - r_i|^3} (\vec{r} - \vec{r}_i)$$

$$= 9 \times 10^9 \times 1 \times 10^{-9} \left[\frac{5 \times 10^{-9} \times (\hat{i} - 3\hat{j} + 3\hat{k})}{\left[\sqrt{1^2 + (-3)^2 + 3^2} \right]^3} + \frac{2 \times 10^{-9} \times (4\hat{i} - 3\hat{j} + 2\hat{k})}{\left[\sqrt{4^2 + (-3)^2 + 2^2} \right]^3} \right]$$

$$= 9 \times 10^9 \left[\frac{5}{19^{3/2}} (-\hat{i} - 3\hat{j} + 3\hat{k}) + \frac{2}{29^{3/2}} (4\hat{i} - 3\hat{j} + 2\hat{k}) \right]$$

$$= 9 \times 10^9 \left[-0.008\hat{i} - 0.2184\hat{j} + 0.2056\hat{k} \right]$$

$$= \left[-0.072\hat{i} - 1.965\hat{j} + 1.85\hat{k} \right] \times 10^{-9} \text{ Newtons}$$

(ii)   Electric field at that point

$$\vec{E} = \frac{\vec{F}}{q} = \frac{\left[-0.072\hat{i} - 1.965\hat{j} + 1.85\hat{k} \right] \times 10^{-9}}{1 \times 10^{-9}}$$

$$= \left[-0.072\hat{i} - 1.965\hat{j} + 1.85\hat{k} \right] \text{ V/m}$$

Q.4 : **If the electric field on a region is** $\vec{E} = 4\hat{i} + 6\hat{j} + 7\hat{k}$, **find the electric flux through the surface area of 75 square units in XY plane.**

Sol.: Surface are in XY plane means that area vector is along z-direction.

So
$$\vec{A} = 75\hat{k}$$

here
$$\vec{E} = 4\hat{i} + 6\hat{j} + 7\hat{k}$$

so electric flux
$$\phi_E = \oint_s \vec{E} \cdot \vec{ds}$$

$$= \vec{E} . \vec{A}$$

$$= \left(4\hat{i} + 6\hat{j} + 7\hat{k}\right) . \left(75\hat{k}\right)$$

$$= 525 \text{ Nm}^2\text{C}^{-1}$$

Q.5 : If 2000 flux lines enter through a given volume of space and 4000 lines diverge from it, calculate the total charge within the volume.

Sol.: Let ϕ_1 = 2000 Vm and ϕ_2 = 4000 Vm.

According to Gauss's law of electrostatics

Total flux $\qquad\qquad \phi = \dfrac{q}{\varepsilon_0} = \phi_2 - \phi_1$

$$q = \varepsilon_0 \left(\phi_2 - \phi_1\right)$$

$$= 8.85 \times 10^{-12} \times 2000$$

$$= 1.77 \times 10^{-8} \text{C} .$$

Q.6 A hollow metallic sphere of radius 0.1 m has 1 × 10⁻⁸C of charge distributed uniformly over it. Calculate electric field intensity (i) on the surface of the sphere (ii) at 8 cm away from the centre and (iii) at point 1 m away from the centre.

Sol.: (i) Electric field intensity on the surface

$$E = \frac{1}{4\pi\varepsilon_0} \frac{q}{r^2}$$

$$= 9 \times 10^9 \times \frac{10^{-8}}{(0.1)^2} = 9 \times 10^3 \text{ N/C} .$$

(ii)   At 8 cm form the centre i.e. point lies inside the sphere.

and E inside sphere = 0

(iii)   and field intensity at 1 m away from the centre

$$E = 9 \times 10^9 \times \frac{10^{-8}}{(1)^2} = 90 \text{ N/C}$$

Q.7 : Two spheres of radii 20 cm and 30 cm are charged separately with 40 esu and 60 esu of charge respectively. Find when they are connected (i) charge will flow from which sphere to which sphere (ii) Common potential.

Sol.: Here V_1 = Potential of sphere of radius 20 cm

$$= \frac{40}{20} = 2\,esu$$

V_2 = Potential of sphere of radius 30 cm

$$= \frac{60}{30} = 2\,esu$$

(i) So charge will not flow.

(ii) When they are connected, at equilibrium the

total charge = 40 + 60 = 100 esu

if the common potential be V then

$$V = \frac{q_1}{c_1} = \frac{q_2}{c_2}$$

so

$$\frac{q_1}{20} = \frac{100 - q_1}{30}$$

$$30\,q_1 = 20 \times 100 - 20\,q_1$$

$$50\,q_1 = 20 \times 100$$

$$q_1 = 40$$

so

$$V = \frac{40}{20} = 2\ esu$$

Q.8: Evaluate if a metal sphere of radius 1 cm hold a charge of 1C?

[WBUT - 2003]

Sol.: As

$$V = \frac{q}{C}$$

But for a spherical capacitor the capacitance

$$C = 4\pi\,\varepsilon_0 r$$

so
$$V = \frac{q}{4\pi\varepsilon_0 r} = 9\times10^9 \times \frac{1}{0.01}$$

$$= 9 \times 10^{11} \text{ Volts}$$

As the breakdown voltage for air is just 3kV/mm hence the surrounding air will be ionized and charge will leak to air.

Q.9 : The spherical region $0 < \gamma < 10$ contains a uniform volume charge density $\rho_V = 4\mu c/m^3$.

(i) **Find $Q_{enclosed}$** $0 < \gamma < 10$ cm.

(ii) **Find D_r.** $0 < \gamma < 10$ cm.   **[RU 2005]**

Sol.: Given

$$\rho_V = 4\mu c/m^3 \qquad\qquad 0 < \gamma < 10 \text{ cm.}$$

from Gauss law

$$\varepsilon_0 \oint \vec{E}.\vec{ds} = Q_{encl.} = \int_V \rho_V dV$$

(i)
$$Q_{encl.} = \int_V \rho_V dV = \rho_V \int dV$$

$$= \rho_V \int_{\phi=0}^{2\pi} \int_{\theta=0}^{\pi} \int_{r=0}^{10} r^2 \sin\theta \, dr \, d\theta \, d\phi$$

$$= \rho_V \frac{4}{3}\pi r^3 = 4\times10^{-6}\times\frac{4}{3}\pi r^3, \qquad 0 < r < 10 \text{ cm}$$

$$= 5.33 \,\pi r^3 \,\mu\, C \qquad 0 < r < 10 \text{ cm}$$

So
$$Q_{encl.} = 16.7456 r^3 \,\mu C$$

(ii)   as
$$\varepsilon_0 \oint_s \vec{E}.\vec{ds} = Q_{encl.}$$

$$D_r \oint_s \vec{ds} = Q_{encl.} = \int_V \rho_V dV = \rho_V \int_V dV$$

$$\Rightarrow \qquad D_r \, 4\pi r^2 = \frac{4\pi r^3}{3} \rho_V$$

$$\Rightarrow \qquad \vec{D} = \frac{r}{3} \rho_V \, \hat{r}, \qquad 0 < r < 10 \text{ cm}$$

$$= \frac{r}{3} \rho_V \, \hat{r}$$

$$= 1.333 \, r \, \mu C \hat{r}, \qquad 0 < r < 10 \text{ cm}.$$

Q.10 : Find the electric flux density $\vec{D}$ at a point A(6, 4, –5) caused by a uniform charge density $\rho_s = 60 \, \mu C / m^2$ at a plane x = 8. **[RU 2003, 2000]**

Sol.: Given that

$$\rho_s = 60 \, \mu C / m^2$$

as

$$\vec{A} = 6\hat{i} + 4\hat{j} - 5\hat{k}$$

$$\vec{D} = \varepsilon_0 \vec{E} = \varepsilon_0 \left(\frac{\rho_s}{2\varepsilon_0} \right) \hat{n}$$

$$= \frac{\rho_s}{2} \hat{n}$$

as plane is x = 8 but point is (6, 4, –5) where x coordinate is less than x = 8 so $\hat{n} = -\hat{i}$

$$\vec{D} = \frac{60 \times 10^{-6}}{2} \left(-\hat{i} \right)$$

$$= -30\hat{i} \, \mu C / m^2 .$$

Q.11 : In cylindrical coordinates $\left(\rho, \phi, z \right)$, electric flux density is given by $\vec{D} = ze \cos^2 \phi \, \hat{z}$ c/m². Calculate the charge density at $\left(1, \frac{\pi}{4}, 3 \right)$ and the total charge enclosed by the cylinder of radius 1 meter with $-2 \le z \le 2$ meter.

[RU 2001, 1999]

Sol.: as charge density is given by

$$\rho_V = \vec{\nabla} \cdot \vec{D} = \frac{\partial \left(z \rho \cos^2 \phi \right)}{\partial z} \, \hat{z} \cdot \hat{z}$$

$$= \rho \cos^2 \phi$$

$$\rho_V \text{ at } \left(1, \frac{\pi}{4}, 3 \right) = \rho \cos^2 \phi$$

$$= 1\cos^2 \frac{\pi}{4}$$

$$= \frac{1}{2} = 0.5$$

and total charge

$$Q = \int_V \rho_V dV = \int_V \rho \cos^2 \phi \, \rho \, d\rho \, d\phi \, dz$$

$$= \int_{z=-2}^{2} dz \int_{\phi=0}^{2\pi} \cos^2 \phi \, d\phi \int_{\rho=0}^{1} \rho^2 \, d\rho$$

$$= \frac{4}{3} \pi \text{ Coulombs}$$

Q.12 : For positive x, y and z, let $\rho_V = 40 \, xyz \, c/m^2$. Calculate the total charge for the region defined by

(i) $0 \le x, y, z \le 2$

(ii) $0, y = 0, 0 \le 2x + 3y \le 10$ and $0 \le z \le 2$ **[RU 2002]**

Sol.: As

$$Q = \int_V \rho_V \, dV$$

(i) $$Q = \int_0^2 \int_0^2 \int_0^2 40 \, xyz \, dx \, dy \, dz$$

$$= 40\left[\frac{x^2}{2}\right]_0^2 \left[\frac{y^2}{2}\right]_0^2 \left[\frac{z^2}{2}\right]_0^2$$

$$= 40 \times 2 \times 2 \times 2 = 320 \text{ coulombs.}$$

(ii)   Again

$$Q = \int\limits_0^5 \int\limits_0^{\frac{10-2x}{3}} \int\limits_0^2 40\ xyz\ dx\ dy\ dz$$

$$= \int\limits_0^5 40x\ dx \left[\frac{y^2}{2}\right]_0^{\frac{10-2x}{3}} \left[\frac{z^2}{2}\right]_0^2$$

$$= \int\limits_0^5 10x\ dx \left(\frac{10-2x}{3}\right)^2$$

$$= \frac{40}{9}\int\limits_0^5 \left(100x - 40x^2 + 4x^3\right)dx$$

$$= \frac{40}{9}\left[100\frac{x^2}{2} - 40\frac{x^3}{3} + 4.\frac{x^4}{4}\right]_0^5$$

$$= \frac{40}{9}\left[50 \times 25 - \frac{40}{3} \times 125 + 625\right]$$

$$= 925.93 \text{ Coulombs.}$$

Q.13 : **Given the potential** $V = \dfrac{10}{r^2}\sin\theta\cos\phi$. **Find electric flux density D at** $\left(2, \dfrac{\pi}{2}, 0\right)$.   **[RU 2006]**

Sol.: We know that

$$\vec{D} = \varepsilon_0\ \vec{E}$$

$$\vec{E} = -\nabla V$$

$$= -\left[\frac{\partial V}{\partial r}\hat{r} + \frac{1}{r}\frac{\partial V}{\partial \theta}\hat{\theta} + \frac{1}{r\sin\theta}\frac{\partial V}{\partial \phi}\hat{\phi}\right]$$

$$= \frac{20}{r^3}\sin\theta\cos\phi\,\hat{r} - \frac{10}{r^3}\cos\theta\cos\phi\,\hat{\theta} + \frac{10}{r^3}\sin\phi\,\hat{\phi}$$

at $\left(2, \dfrac{\pi}{2}, 0\right)$

$$\vec{D} = \varepsilon_0\,\vec{E}$$

$$= \varepsilon_0\left[\frac{20}{r^3}\hat{r} - \frac{10}{r^3}.0\,\hat{\theta} + \frac{10}{r^3}.0.\hat{\phi}\right]$$

as $r = 2$

$$D = 2.5\,\varepsilon_0\,\hat{r}$$

$$= 22.12\times10^{-12}\hat{r}\ \text{c/m}$$

Q.14 : **Given the potential** $V = \dfrac{10}{r^2}\sin\theta\cos\phi$**. Find the work done in moving a**

$10\ \mu c$ **charge from point A (1, 30°, 120°) to B(4, 90°, 60°).**

Sol.: The work done

$$W = -Q\int_{A}^{B}E\cdot dl = QV_{AB}$$

or $W = Q\left(V_B - V_A\right)$

$$= 10\left[\frac{10}{16}\sin 90°\cos 60° - \frac{10}{1}\sin 30°\cos 120°\right]\times10^{-6}$$

$$= 10\left[\frac{10}{32} - \frac{-5}{2}\right]\times10^{-6}$$

$$= 28.125\times10^{-6}\ \text{Joule}$$

Q.15 : Calculate electric field $\vec{E}$ at (20, 0) and (4, 6) for a potential value of V = $10y^3 + 20x^2$.

Sol.: As electric is given by

$$\vec{E} = -\vec{\nabla}V$$

$$= -\left(\frac{\partial}{\partial x}\hat{i} + \frac{\partial}{\partial y}\hat{j}\right)\left(10y^3 + 20x^2\right)$$

$$= -\left[\frac{\partial}{\partial x}\hat{i}\left(10y^3 + 20x^2\right) + \frac{\partial}{\partial y}\hat{j}\left(10y^3 + 20x^2\right)\right]$$

$$= -40x\,\hat{i} - 30y^2\hat{j}$$

Now at (20, 0)

$$E_{20,\,0} = -800\,\hat{i}$$

and at (4, 6)

$$E_{4,\,6} = -160\hat{i} - 108\hat{j}$$

Q.16 : Calculate the value of potential at any point inside and outside a uniformly charged sphere of radius 'R'.

Sol.:

Consider a sphere of radius 'R' and any point of consideration P at any distance r. Assume ρ_V be the volume charge density.

When r > R

$$\vec{E} = \frac{Q}{4\pi\varepsilon_0 r^2}\hat{r} = \frac{\rho_V \int_V dV}{4\pi\,\varepsilon_0 r^2}\hat{r}$$

$$= \frac{\rho_V \left(\frac{4}{3}\pi R^3\right)}{4\pi\varepsilon_0 r^2}\hat{r}$$

$$= \frac{\rho_V R^3}{3\,\varepsilon_0 r^2}\hat{r}$$

and electric potential

$$V = -\int\limits_0^r \vec{E} \cdot \vec{dl}$$

$$= -\int\limits_0^r \frac{\rho_V R^3}{3\,\varepsilon_0 r^2}\,(\hat{r}.\,\hat{r})\,dr \qquad\qquad \left[\text{as } dl = \hat{r}\,dr\right]$$

$$= \frac{\rho_V\,R^3}{3\varepsilon_0\,r}$$

when r < R

$$E = \frac{Q}{4\pi\varepsilon_0 r^2}\,\hat{r} \quad (0 < r < R)$$

$$= \frac{\rho_V\left(\dfrac{4}{3}\pi\,r^3\right)}{4\pi\varepsilon_0 r^2}$$

$$= \frac{\rho_V r}{3\varepsilon_0}$$

Electric potential $$V = -\int \vec{E} \cdot \vec{dl}$$

$$= \left[\int\limits_0^R \frac{\rho_V R^3}{3\varepsilon_0 r}(\hat{r}.\hat{r})\,dr - \int\limits_R^r \frac{\rho_V r(\hat{r}.\hat{r})}{3\,\varepsilon_0}\,(\hat{r}.\hat{r})\,dr\right]$$

$$= \frac{\rho_V R^2}{3\,\varepsilon_0} - \frac{\rho_V r^2}{6\varepsilon_0} + \frac{\rho_V R^2}{3\varepsilon_0}$$

$$= \frac{\rho_V R^2}{3\,\varepsilon_0}\left[3R^2 - r^2\right]$$

Q.17 : When electric field is always directed towards X-direction then prove that potential will be independent of Y and Z coordinates. Also prove that when field is constant, there will not be any free charge in that region.

[WBUT - 2007]

Sol.: According to the questions

$$\vec{E} = \vec{E}_x \hat{i} \qquad \qquad ...(1)$$

and

$$\vec{E} = -\vec{\nabla} V$$

$$= -\frac{\partial V_x}{\partial x}\hat{i} - \frac{\partial V_y}{\partial y}\hat{j} - \frac{\partial V_z}{\partial z}\hat{k} \qquad \qquad ...(2)$$

Comparing (1) and (2)

$$\frac{\partial V_y}{\partial y} = 0 \quad \text{and} \quad \frac{\partial V_z}{\partial z} = 0$$

so $V_y = V_z = \text{constant}$

Again $\vec{E} = -\vec{\nabla} V$ shows that when

$$\vec{E} = \text{constant} \rightarrow \vec{\nabla} V = \text{constant}$$

So $V(x,y,z) = 0$

and form poisson's relation

$$\nabla^2 V = \frac{\rho}{\varepsilon_0}$$

or $\rho = \varepsilon_0 \nabla^2 V = 0$

i.e. it is charge free region.

Q.18 : A long cylinder carries charge proportional to the distance form the axis (r). If the cylinder is with radius R, then find the electric field both at r > R and r < R using Gauss's law of electrostatics.

Sol.: Consider a long cylinder of radius R and height h.

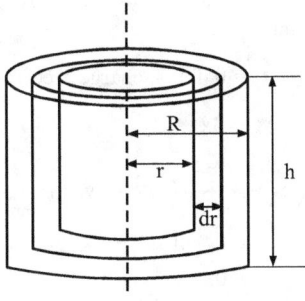

so the charge enclosed

$$Q = \int\limits_0^R (kr)(2\pi \, rdr)h$$

$$= \frac{k \, 2\pi h. \, R^3}{3}$$

Case - I : if $r > R$ then

$$\int \vec{E}.\,\overline{ds} = \frac{Q}{\varepsilon_0}$$

or $\qquad\qquad E \, 2\pi \, rh = \dfrac{k. \, 2\pi h. \, R^3}{3\varepsilon_0}$

$$E = \frac{k \, R^3}{3\varepsilon_0 r}$$

Case - II : When $r < R$ then

$$q' = \int\limits_0^h (kr)(2\pi \, rh)\,dr$$

$$= \frac{k. 2\pi h \, r^3}{3}$$

so $\qquad\qquad E = \dfrac{kr^3}{3\varepsilon_0 r} = \dfrac{kr^2}{3\varepsilon_0}$

Q.19 : Verify that potential function $V_{x,y} = 3x^2 + 2y^2$ satisfies Laplace's equation or not. Also find the charge density.

Sol.: Laplace's equations states that

$$\nabla^2 V = 0 \text{ for a charge free region}$$

here $\qquad\qquad V = 3x^2 + 2y^2$

$$\nabla^2 V = \frac{\partial^2}{\partial x^2}\left(3x^2 + 2y^2\right) + \frac{\partial^2}{\partial y^2}\left(3x^2 + 2y^2\right)$$

$$= 6 + 4 = 10$$

$$\neq 0$$

as $\nabla^2 V \neq 0$ it shows that potential function does not satisfy Laplace equation and because it is not a charge free region here poisson's equation will hold true

where $$\nabla^2 V = -\frac{\rho}{\varepsilon_0}$$

here $$-\frac{\rho}{\varepsilon_0} = 10$$

so $$\rho = -10\,\varepsilon_0 \text{ C/m}^3$$

Q.20 : A condenser consisting of an oxidized aluminimum sheets with effective surface area of 400 cm² with a capacitance of 8μF. Calculate

(i) The field strength if $\varepsilon_r = 8$ and potential difference between aluminium and the electrolyte = 10 V.

(ii) Total dipole moment induced in the oxide layer and its susceptibility.

Sol.: Here

$$A = 400 \text{ cm}^2 = 400 \times 10^{-4} \text{ m}^2$$

$$\varepsilon_r = 8, \ C = 8 \ \mu F, \ V = 10 \text{ V}$$

Electric field $$E = \frac{V}{d}$$

and $$d = \frac{\varepsilon_0\,\varepsilon_r A}{C}$$

$$= \frac{8.854 \times 10^{-12} \times 8 \times 400 \times 10^{-4}}{8 \times 10^{-6}}$$

$$= 3541 \times 10^{-10} \text{ m.}$$

(i) $$E = \frac{V}{d} = \frac{10}{3541 \times 10^{-10}} = 2.8 \times 10^7 \text{ V/m}$$

(ii) Polarization $$P = \varepsilon_0 E\,(\varepsilon_r - 1)$$

$$= 8.854 \times 10^{-12} \times 2.8 \times 10^7\,(8-1)$$

$$= 1.7354 \times 10^{-3}$$

and
$$P = \frac{p}{volume}$$

so
$$p = P \times volume = P \times (A \times d)$$
$$= 1.7354 \times 10^{-3} \times 400 \times 10^{-4} \times 3541 \times 10^{-10}$$
$$= 2.5 \times 10^{-11}$$

and susceptibility

$$\chi = \frac{p}{\varepsilon_0 E}$$

$$= \frac{2.5 \times 10^{-10}}{8.854 \times 10^{-12} \times 2.8 \times 10^7}$$

$$= 0.1 \times 10^{-6}$$

Q.21 : A dielectric material contains 2×10^9 polar molecules/m³ each of dipole moment 1.8×10^{-27} Cm. Assuming that all of the dipoles are aligned towards electric field $E = 10^5$ V/m. Find the polarization, electric susceptibility and the relative permittivity.

[IES 97]

Sol.: It is given that

No. of molecules $N = 2 \times 10^9$ moleucles/m³

Dipole moment $p = 1.8 \times 10^{-27}$ c-m

Electric field $E = 10^5$ V/m.

Hence polarization

$$P = Np$$
$$= 2 \times 10^9 \times 1.8 \times 10^{-27}$$
$$= 3.6 \times 10^{-18} \text{ C/m}^2$$

susceptibility

$$\chi = \frac{P}{\varepsilon_0 E} = \frac{3.6 \times 10^{-18}}{8.854 \times 10^{-12} \times 10^5}$$
$$= 4.07 \times 10^{-12}$$

and relative permittivity

$$\varepsilon_r = 1 + \chi = 1 + 4.07 \times 10^{-12}$$

$$\approx 1$$

Q.22 : A certain homogenous slab of loss-less dielectric material is characterized by an electric susceptibility of 0.12 and carries a uniform flux density within it of 1.6 n C/m². Calculate the electric field intensity, the polarization, and the average dipole moment. Given No. of dipoles per cubic meter = 2 × 10¹⁹ and separation between two equipotential surfaces = 2.54 cm.　　　　　　　　　　　　　　　**[IES - 1998]**

Sol.: Electrical susceptibility

$$\chi_e = 0.12, \ D = 1.6 \text{ n C/m}^2$$

We know

$$\varepsilon_r = 1 + \chi_e$$

$$= 1 + 0.12$$

$$= 1.12$$

$$D = \varepsilon_0 \ \varepsilon_r E$$

$$E = \frac{D}{\varepsilon_0 \ \varepsilon_r} = \frac{1.6 \times 10^{-9}}{8.854 \times 10^{-12} \times 1.12}$$

$$= 161.34 \text{ V/m}$$

and voltage

$$V = E \times \text{ separation}$$

$$= 161.34 \times 2.54 \times 10^{-2}$$

$$= 4$$

Polarization $P = \chi_e \ \varepsilon_0 E$

$$= 0.12 \times 8.854 \times 10^{-12} \times 161.34$$

$$= 0.1714 \text{ n C/m}^2$$

and　Average dipole moment $= \dfrac{0.1714 \times 10^{-9}}{2 \times 10^{19}}$

$$= 8.8714 \times 10^{-30} \text{ C-m.}$$

Summary

- Vector form of Coulomb's law

$$F = \frac{1}{4\pi\varepsilon_0} \frac{q_1 q_2}{|\vec{r}|^2} \hat{r}$$

- Linear charge density

$$\lambda = \lim_{\Delta l \to 0} \frac{\Delta q}{\Delta l} \ C/m$$

surface charge density

$$\sigma = \lim_{\Delta s \to 0} \frac{\Delta q}{\Delta s} \ C/m^2$$

Volume charge density

$$l = \lim_{\Delta V \to 0} \frac{\Delta q}{\Delta V} \ C/m^3$$

- Electric field strength

$$\vec{E} = \frac{1}{4\pi\varepsilon_0} \frac{q}{r^2} \hat{r} \ N/C$$

- Electric displacement vector

$$\vec{D} = \varepsilon_m \vec{E}$$

- Electrostatic potential energy

$$U = \frac{1}{4\pi\varepsilon_0} \frac{q_1 q_2}{r_{12}}$$

- Electrostatic potential is electrostatic potential energy per unit positive test charge

$$V = \frac{U}{q_2} = \frac{1}{4\pi\varepsilon_0} \frac{q_1}{r_{12}}$$

- Electric flux

$$\phi = \oint_s \vec{E} \cdot \vec{ds}$$

- Gauss law of electrostatic

$$\oint_s \vec{E} \cdot \vec{ds} = \frac{Q_{enclosed}}{\varepsilon_0}$$

or
$$\vec{D} \oint_s \vec{ds} = Q_{enclosed}$$

- Differential form of Gauss's law

$$\vec{\nabla} . \vec{E} = \frac{\rho}{\varepsilon_0}$$

or
$$\vec{\nabla} . \vec{E} = \rho$$

- Electric field $\qquad \vec{E} = -\vec{\nabla} V$
- Poisson's equation

$$\nabla^2 V = \frac{-\rho}{\varepsilon_0}$$

- Laplace equation for charge free region

$$\nabla^2 V = 0$$

- Equation of continuity

$$\vec{\nabla} . \vec{J} + \frac{\partial \rho}{\partial t} = 0$$

- Drift velocity

$$v_d = \frac{eE}{m} \tau \text{ where } \tau \text{ is relaxation time.}$$

- Electric current

$$I = ne A vd$$

- Current density

$$\vec{J} = \sigma \vec{E}$$

- Dipole moment

$$\vec{p} = q \vec{r}$$

- Polarization $\qquad \vec{P} = \chi \varepsilon_0 \vec{E}$

- $\qquad\qquad\qquad \vec{D} = \varepsilon_0 \vec{E} + \vec{P}$

- Dielectric constant or relative permittivity

$$\varepsilon_r = k = 1 + \chi$$

EXERCISE

1. Give statement of coulomb's law of electrostatics. Calculate the electric field intensity and electric potential for any charge.

2. What do you understand by electric flux and velocity flux?

3. Give statement of Gauss' law of electrostatic. Give its differential form also.

4. Using Gauss's law of electrostates, derive Coulomb's law.

5. What is electrostatic potential energy and electrostatic potential?

6. From the differential form of Gauss' law, find poisson's and Laplace's equations.

7. For any electric field $\vec{E}$ prove that $\vec{\nabla} \times E = 0$

8. For any electric field prove that $\vec{E} = -\vec{\nabla}V$ where V is potential.

9. Find electric field intensity at any point inside and outside for a cylindrical charge distribution.

10. Explain the behaviour of a dielectric material placed in an electric field and hence explain electric polarization of matter.

11. What are polar and non-polar molecules? Discuss different types of polarizations in dielectrics.

12. What do you understand by atomic polarizability? Find a relation between dipole moment and atomic polarizability.

13. Explain the terms permittivity, dielectric coefficients, susceptibility and dielectric polarization. Derive their relation also.

14. Show that $\vec{D} = \varepsilon_0 \vec{E} + \vec{P}$

15. Define the terms dielectric constant and electrical susceptibility and prove that
    $$k = 1 + \chi_e.$$

16. State and prove Gauss's law in dielectrics.

17. For any electric potential given by $V(x,y,z) = \left(2x^2 + 4y^2 + 3z^2\right)$ find the electric field at (1, 1, 1).   [WBUT 2005]

18. In space the electric field is given by $\vec{E} = 8\hat{i} + 4\hat{j} + 3\hat{k}$. Calculate the electric flux through a surface area 100 sq. units in xy planes.   [WBUT 2004]

3

MAGNETOSTATICS AND TIME VARYING FIELDS

Contents: Lorentz forces, force on a small current element placed in a magnetic field, Biot-Savart-law and its applications, divergence of magnetic field, vector potential, Ampere's law in integral form and conversion to differential form. Faraday's law of electromagnetic induction in integral form and conversion to differential form.

FORCE ON A CURRENT CARRYING CONDUCTOR IN A MAGNETIC FIELD

When a current carrying conductor is placed in a magnetic field, a magnetic force works on it in perpendicular direction to both current and magnetic field.

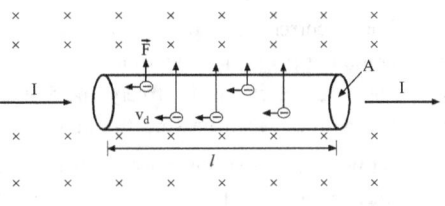

Consider a portion of length l and cross sectional area A of a conductor carries a current I placed in any magnetic field $\vec{B}$ directed downward into the page.

As current in the conductor is due to free electron drift from lower to higher potential ends of the conductor, magnetic force works on these electrons. If there are n electrons per unit volume then from

$$F = q\,\vec{v} \times \vec{B}$$

for each electrons $\qquad F' = e\,v_d B \qquad\qquad\qquad\qquad\qquad …(i)$

No. of electrons for 1 length

$$N = n\,Al$$

so total force

$$F = F'N$$

$$= (e\,v_d B)(n\,Al)$$

$$= (n\,e\,Av_d)Bl$$

$$F = I\,Bl \qquad\qquad (\text{as } I = neAvb_d)$$

if θ is the angle between conductor and $\vec{B}$ then

$$\boxed{F = IBl\,\sin\theta} \qquad\qquad ...(ii)$$

or

$$\boxed{\vec{F} = I\,\vec{l} \times \vec{B}} \qquad\qquad ...(iii)$$

where $\vec{l}$ is a displacement vector in the direction of current.

LORENTZ FORM

It has been observed that stationary electric charge does not have any effect on the nearby magnet shown no evidence of any magnetic field when charge is at rest, but H.C. Oersted found that when any wire with current in it is placed near to any magnetic compass, the needle of compass deflects as soon as their is the movement of charge, showing presence of magnetic field when there is current i.e. charge is in motion. A compass needle experiences a force, when placed near to a current carrying conductor.

Since a current has moving electric charges, that means moving electric charge is responsible for generation of magnetic field around the conductor causing compass needle to deflect. It clearly shows that free electrons or freely moving charged practices (not in the wire) would also experience a force when passing through a magnetic field.

Let us assume that there are N charged particles of charge q pass by a given point in time t, then there is a current

$$I = \frac{Nq}{t}$$

Suppose t be the time for a charge to travel a distance l in presence of a magnetic field $\vec{B}$. Then

$$\vec{l} = \vec{v}\,t, \text{ where } \vec{v} \text{ is the velocity of charged particle.}$$

The force on these N particles

$$\vec{F} = I\vec{l} \times \vec{B} = \left(\frac{Nq}{t}\right)(\vec{v}t) \times \vec{B}$$

$$= Nq\,\vec{v} \times \vec{B}$$

so force on any one particle is

$$\vec{F} = q\,\vec{v} \times \vec{B} \qquad \qquad ...(i)$$

when an electric charge is at rest or move, in any electrostatic field $\vec{E}$, then the electrostatic

force $= q\vec{E}$ ....(ii)

This combined effect of force experienced by a charged particle moving in an electric and magnetic field is known as lorentz forces.

so if charge q moves with a velocity $\vec{v}$ in a combined electric and magnetic field, then the total force, known as Lorentz force is given by

$$\vec{F} = q\,\vec{E} + q\left(\vec{v} \times \vec{B}\right)$$

$$\boxed{\vec{F} = q\left[\vec{E} + \vec{v} \times \vec{B}\right]} \qquad ...(iii)$$

The above equation is also known as Lorentz equation for total electromagnetic force on any charge moving in an electric field along with a magnetic field.

Here magnetic Lorentz force varies directly with the strength of magnetic field $\vec{B}$ as well as with velocity $\vec{v}$. So if charged particle stops i.e. $v = 0$ so magnetic lorentz force becomes zero. the magnetic Lorentz force does not work on charge because its direction is normal to the charge motion. As $F = q\left(\vec{v} \times \vec{B}\right)$ which shows that Lorentz magnetic force is in the direction of $\vec{v} \times \vec{B}$ i.e. perpendicular to both direction of charge motion $\left(\vec{v}\right)$ and strength of magnetic field.

SOME IMPORTANT TERMS

Magnetic flux

Magnetic flux is defined as the total number of magnetic field lines passing through a surface in the normal direction.

Mathematically it is equal to

$$\phi_B = \oint_s \vec{B}.\,\overrightarrow{ds} \qquad\qquad \text{Its unit is weber (wb)}$$

Magnetic flux density $(\vec{B})$ or magnetic induction $(\vec{B})$

It is defined as the magnetic flux per unit area. Its SI unit in wb/m² or Tesla (T).

Magnetic Intensity $(\vec{H})$

It is defined as the ratio of magnetic flux density $(\vec{B})$ to the permeability (μ) of the medium.

so
$$\vec{H} = \frac{\vec{B}}{\mu}$$

Its SI unit is A/m.

Biot-Savart Law

Experimentally it has been verified that a current carrying conductor can produce a magnetic field around it. The direction and magnitude of this magnetic field can be determined using Biot-Savrt law, according to which "the magnetic induction $d\vec{B}$ at any point P due to an element of infinitesimal length $\vec{dl}$ with current I is proportional to (a) the current element dl. (b) the current I through dl (c) sine of angle θ between the element dl and radius vector r. and (d) inversely proportional to the radius vector r which is the distance of point P from dl.

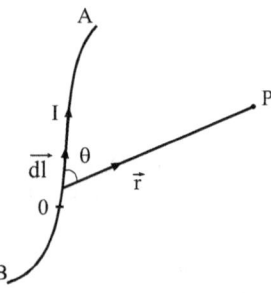

i.e.
$$dB \propto \frac{I\,dl\,\sin\theta}{r^2}$$

$$d\vec{B} = \frac{\mu_0}{4\pi}\frac{I\,dl\,\sin\theta}{r^2} \qquad \text{...(i)}$$

where μ_0 is the permeability of free space.

and
$$\mu_0 = 4\pi \times 10^{-7} \text{ wb A}^{-1}\text{m}^{-1}$$

$$\vec{dB} = \frac{\mu_0}{4\pi}I\frac{\vec{dl}\times\vec{r}}{r^3} = \frac{\mu_0}{4\pi}I\frac{\vec{dl}\times\hat{r}}{r^2} \qquad \text{...(ii)}$$

equation (ii) is the vector form of Bio-Savart law.

Now if

$$\theta = 0° \rightarrow \left|d\vec{B}\right| = 0$$

which shows that magnitude of magnetic flux density becomes zero at any axial point on the current carrying conductor.

if $\theta = 90°$ then

$$\left|d\vec{B}\right| = \frac{\mu_0}{4\pi} \frac{Idl}{r^2}$$

it shows the maximum value of magnetic field intensity at any point on perpendicular line to the current carrying length element.

The total magnetic field intensity due to complete length of current carrying conductor

$$\vec{B} = \int d\vec{B}$$

$$\vec{B}(\vec{r}) = \frac{\mu_0 I}{4\pi} \int \frac{\vec{dl} \times \hat{r}}{r^2}$$

$$\boxed{\vec{B}(\vec{r}) = \frac{\mu_0 I}{4\pi} \int \frac{\vec{dl} \times \vec{r}}{r^3}} \qquad ...(iii)$$

Equation (iii) is the integral form of biot-Savart's law which is analogous to Coulomb's law of electrostatics. The current element I dl can be represented in terms of surface current density J_s or volume current density J_v as

$$I \vec{dl} = J_s \ ds$$

and $$I \ dl = J_v \ dV$$

and hence Biot-Savart's law can accordingly be written as

$$\vec{B} = \frac{\mu_0}{4\pi} \int_s \frac{\vec{J}_s \times \vec{r}}{r^3} \ \overrightarrow{dS} \quad \text{(surface current distribution)} \qquad ...(iv)$$

and for volume current distribution

$$\vec{B} = \frac{\mu_0}{4\pi} \int_V \frac{\vec{J}_v \times \vec{r}}{r^3} \ \overrightarrow{dV} \qquad ...(v)$$

APPLICATIONS OF BIOT-SAVRT'S LAW

(i)   Magnetic field at the centre of a current carrying circular coil

Suppose there is a circular coil of radius r, carrying a current I as shown

According to Biot-Savart law, the magnitude of magnetic field at O due to a small element dl of the coil is

here

$$dB = \frac{\mu_0}{4\pi} \frac{I\,dl\,\sin\theta}{r^2}$$

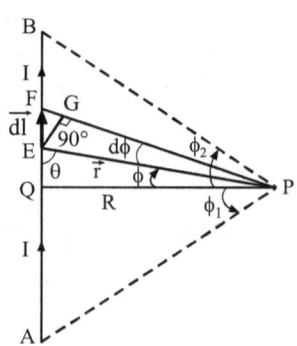

But $\theta = 90°$

so $dB = \frac{\mu_0}{4\pi} \frac{I\,dl}{r^2}$

and total field due to entire coil

$$B = \frac{\mu_0}{4\pi} \frac{I}{r^2} \int dl$$

$$= \frac{\mu_0}{4\pi} \frac{I}{r^2} (2\pi r) \left(\text{as } \int dl = 2\pi r \right)$$

$$\boxed{B = \frac{\mu_0}{2} \frac{I}{r}}$$

Magnetic field B is perpendicular to the plane of the coil, directed upwards. If the current is reversed the field will be downward directed.

(ii)   Magnetic field due to a straight current carrying conductor of finite length

Suppose AB is a straight conductor carrying a current of I and magnetic field intensity is to be determined at point P.

According to Biot-Savart law the magnetic field at P

$$\overline{dB} = \frac{\mu_0}{4\pi} \frac{I \overrightarrow{dl} \times \overrightarrow{r}}{r^3}$$

angle between $I\,\overrightarrow{dl}$ and $\overrightarrow{r}$ is $(180 - \theta)$ so

$$dB = \frac{\mu_0}{4\pi} \frac{I\,dl\,\sin(180-\theta)}{r^2}$$

$$dB = \frac{\mu_0}{4\pi} \frac{I\,dl\,\sin\theta}{r^2} \qquad ...(i)$$

Now $\quad$ EG = EF sin θ

$$= dl\ sin\ θ$$

and $\quad$ EG = EP sin dφ = r sin dφ

$$= r\ dφ$$

so $\quad$ dl sin θ = r dφ $\qquad$ ...(ii)

so from (i)

$$dB = \frac{μ_0}{4π}\frac{I\,dφ}{r} \qquad ...(iii)$$

from Δ EQP, $\qquad r = \dfrac{R}{\cos φ}$

so $\qquad dB = \dfrac{μ_0}{4π}\dfrac{I\cos φ\,dφ}{R} \qquad ...(iv)$

Then the total magnetic field at point P due to the entire conductor is

$$B = \int_{-φ_1}^{φ_2} \frac{μ_0}{4π}\frac{I}{R}\cos φ\,dφ$$

$$= \frac{μ_0}{4π}\frac{I}{R}\left[\sin φ\right]_{-φ_1}^{φ_2}$$

$$\boxed{B = \frac{μ_0}{4π}\frac{I}{R}\left(\sin φ_1 + \sin φ_2\right)} \qquad ...(v)$$

for any conductor of infinite length

$$φ_1 = φ_2 = 90°$$

so $\qquad B = \dfrac{μ_0}{4π}\dfrac{2I}{R}$

$$\boxed{B = \frac{μ_0}{2π}\frac{I}{R}}\ NA^{-1}\ m^{-1}$$

The direction of magnetic field due to a current carrying conductor can be obtained by using any of the laws like

    (i)   Right hand palm rule no 1

    (ii)   Right hand thumb rule or

    (iii) Maxwell Right-hand screw Rule.

(iii) Force between two parallel current carrying conductors and definition of ampere

Experimentally it has been observed that two current carrying conductors attract each other when the currents in them are in same direction and repel each other when these are in opposite directions.

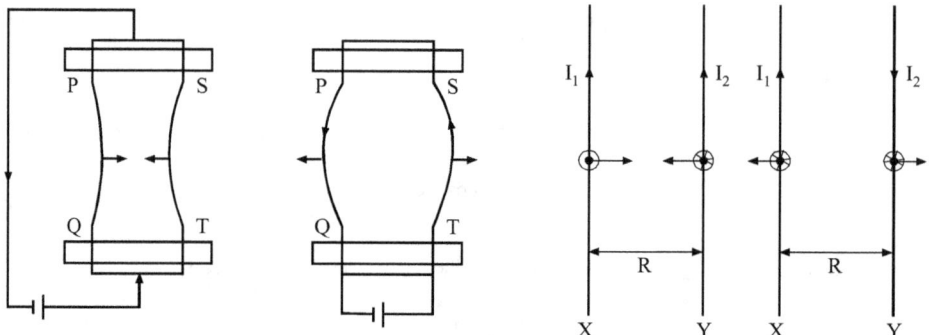

Suppose X and Y are two long parallel straight conductors with currents I_1 and I_2 amp. respectively. The magnitude of the magnetic field $\vec{B}$, at any point on Y due to the current I_1 in X is given by

$$B_1 = \frac{\mu_0\, I_1}{2\pi\, R}$$

Perpendicular to plane of the page directed downward.

Hence current carrying conductor Y is thus situated in a magnetic field $\vec{B}_1$ perpendicular to its length. Hence it experiences a magnetic force, magnitude of which is given by

$$F = I_2\, B_1 l$$

$$= I_2 \left(\frac{\mu_0 I_1}{2\pi R} \right) l$$

and force per unit length

$$\boxed{\frac{F}{l} = \frac{\mu_0}{2\pi}\, \frac{I_1 I_2}{R}}$$

according to Fleming left hand rule the direction of this force is towards X when I_1 & I_2 are in same direction and is away from X when I_1 & I_2 are in opposite direction.

Similarly force per unit length of X due to current in Y is $\dfrac{\mu_0}{2\pi}\dfrac{I_1 I_2}{R}$, directed opposite to forces on Y due to X.

The directions of the forces on the two conductors show that the conductors attract each other if currents are in same direction and repel each other if currents are in opposite direction.

If $I_1 = I_2 = I$

so $\dfrac{F}{1} = \dfrac{\mu_0}{2\pi}\dfrac{I^2}{R}$

If now the value of current is such that when R = 1 m, $\dfrac{F}{1} = 2 \times 10^{-7}$ N/m, then the current is said to be one ampere. So one ampere is that current which can produce a force of 2×10^{-7} N/m in vacuum between two infinitely long conductors placed 1 meter apart.

So $2 \times 10^{-7}\ \dfrac{N}{m} = \dfrac{\mu_0}{2\pi}\dfrac{(1A)^2}{1m}$

so $\dfrac{\mu_0}{4\pi} = 10^{-7}\ \dfrac{N}{A^2}\left[= 10^{-7}\ \dfrac{Wb}{A-m}\right]$

(iv) Magnetic field at the axis of a current carrying circular coil

Suppose there is a circular coil of radius a carrying a current I

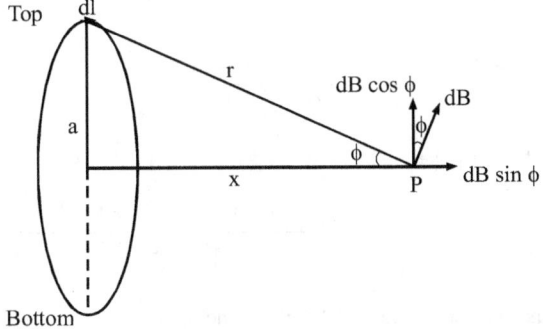

According to Biot-Savart law, the magnitude of magnetic field due to the small current element at P will be

$$dB = \frac{\mu_0}{4\pi} \frac{I \, dl \sin \theta}{r^2}$$

here θ is the angle between length element and the line joining to the point P. i.e. 90°.

so $$dB = \frac{\mu_0}{4\pi} \frac{Idl}{r^2}$$ ....(i)

The direction will be right angle to the line joining length element to point P. This can be resolved into two components dB sin ϕ along the axis of coil and other dB cos ϕ at right angle to axis. The components which are along the axis will be added up. All the vertical components are equal and opposite and cancel each other.

Hence the resultant magnetic field B at P is directed along the axis is

$$B = \int dB \sin \phi$$

$$= \frac{\mu_0 I}{4\pi r^2} \int dl \sin \phi$$

here $$\sin \phi = \frac{a}{r}$$

so $$B = \frac{\mu_0}{4\pi} \frac{Ia}{r^3} \int dl$$

But $$\int dl = 2\pi a$$

and $$r^2 = a^2 + x^2$$

so $$B = \frac{\mu_0}{4\pi} \frac{2\pi \, Ia^2}{\left(a^2 + x^2\right)^{3/2}}$$

$$= \frac{\mu_0 \, Ia^2}{2\left(a^2 + x^2\right)^{3/2}}$$

if the coil has N turns then

$$B = \frac{\mu_0}{4\pi} \frac{2\pi N Ia^2}{\left(a^2 + x^2\right)^{3/2}} = \frac{\mu_0 N Ia^2}{2\left(a^2 + x^2\right)^{3/2}} NA^{-1}m^{-1}$$

direction of this magnetic field will be along the axis.

klif x = 0 i.e. at the centre of coil

$$B = \frac{\mu_0}{4\pi} \frac{2\pi\, NI}{a}$$

$$B = \frac{\mu_0 NI}{2a} \; NA^{-1}m^{-1}$$

(v)   Magnetic field of a solenoid

A solenoid is considered to be a long cylindrical helix.

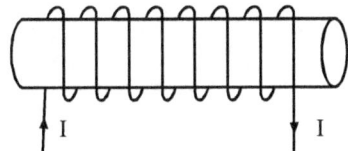 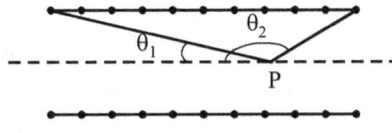

Figure shows a solenoid of radius a meter and carrying a current I ampere. The lines of forces inside the solenoid are nearly parallel which clearly represents that the magnetic field within the solenoid is uniform and parallel to the axis of solenoid.

Suppose there are n turns per unit length of the solenoid. Consider solenoid to be divided up into a number of small coils and consider one of those coils is AB of dx width.

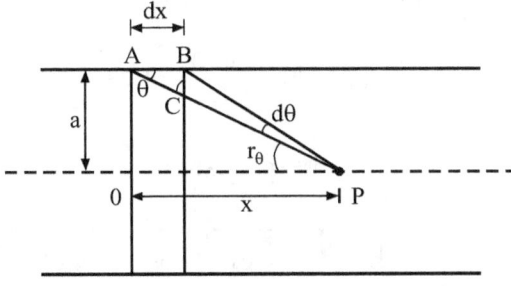

Fig.

The number of turns in this coil is n dx.

The magnetic field at P due to this small coil is

$$dB = \frac{\mu_0 (n\, dx) I\, a^2}{2(a^2 + x^2)^{3/2}} \; NA^{-1}m^{-1}$$

according to the figure

$$\sin \theta = \frac{BC}{AB} = \frac{rd\theta}{dx}$$

or

$$dx = \frac{rd\theta}{\sin \theta}$$

and

$$r^2 = a^2 + x^2$$

Hence

$$dB = \frac{\mu_0 (nr \, d\theta) \, Ia^2}{2r^3 \sin \theta}$$

$$= \frac{\mu_0 n Ia^2}{2r^2} \frac{d\theta}{\sin \theta}$$

and

$$\sin \theta = \frac{a}{r}$$

or

$$a^2 = r^2 \sin^2 \theta$$

so

$$dB = \frac{\mu_0 n I}{2} \sin \theta \, d\theta$$

The total magnetic field at P can be obtained between the first and last turn of the solenoid i.e. between the limit of θ_1 and θ_2.

so

$$B = \int_{\theta_1}^{\theta_2} dB = \int_{\theta_1}^{\theta_2} \frac{\mu_0 n I}{2} \sin \theta \, d\theta$$

$$= \frac{\mu_0 n I}{2} \left[-\cos \theta \right]_{\theta_1}^{\theta_2}$$

$$\boxed{B = \frac{\mu_0 n I}{2} \left(\cos \theta_1 - \cos \theta_2 \right)} \quad NA^{-1} \, m^{-1}$$

If P lies will inside the solenoid

i.e. $\theta_1 \approx 0$ and $\theta_2 \approx 180°$ so

$$B = \frac{\mu_0 n I}{2} \left[1 - (-1) \right]$$

$$\boxed{B = \mu_0 n I} \quad NA^{-1} \, m^{-1}$$

at the ends of the solenoid

$$\theta_1 = 0$$

and $$\theta_2 = 90°$$

so $$\boxed{B = \frac{\mu_0 nI}{2}}\ NA^{-1}m^{-1}$$

which clearly shows that the magnetic field at the ends of a solenoid is half of that at the centre of the solenoid.

for a very long solenoid, field is almost uniform and is parallel to the solenoidal axis.

Gauss's law in magnetism (Divergence of magnetic field)

We know that divergence of any vector field can be defined as the limiting value of ratio of flux across any closed surface around the point to the volume when volume is contracted to zero.

i.e. $$\vec{\nabla}\cdot\vec{B} = \lim_{\Delta V \to 0}\ \frac{\oint_s \vec{B}\cdot\overrightarrow{ds}}{\Delta V} \qquad ...(i)$$

The magnetic flux through a closed surface is always zero

i.e. $$\oint_s \vec{B}\cdot\overrightarrow{ds} = 0 \qquad ...(ii)$$

which is Gauss's law in magnetism.

From Gauss's Divergence theorem

$$\oint_s \vec{B}\cdot\overrightarrow{ds} = \int_V \left(\vec{\nabla}\cdot\vec{B}\right) dV \qquad ...(iii)$$

from (ii) & (iii)

$$\int_V \left(\vec{\nabla}\cdot\vec{B}\right) dV = 0 \qquad ...(iv)$$

Hence $$\vec{\nabla}\cdot\vec{B} = 0 \qquad ...(v)$$

Equation (v) shows that free isolated magnetic poles does not exist i.e. magnetic poles always exist in pairs.

Vector Potential

If the divergence of any vector is zero, then there exists a vector $\vec{A}$ whose curl gives the given vector. Such a vector is called vector potential of that vector field.

As $\text{div}\left(\text{curl } \vec{A}\right) = 0$ then $\vec{A}$ is the vector field.

In magnetostatic, the divergence of magnetic field induction $\vec{B}$ is zero,

i.e., $\qquad\qquad\qquad\qquad \vec{\nabla} \cdot \vec{B} = 0 \qquad\qquad\qquad\qquad\qquad$ ...(i)

and $\qquad\qquad \vec{\nabla} \cdot \left(\vec{\nabla} \times \vec{A}\right) = 0 \qquad\qquad\qquad\qquad\qquad$ ...(ii)

By comparing (i) and (ii)

$$\vec{B} = \vec{\nabla} \times \vec{A} \qquad\qquad\qquad ...(iii)$$

hence here $\vec{A}$ can be the vector potential of magnetic field $\vec{B}$.

Hence any magnetic vector potential is defined as a vector function, the curl of which gives the magnetic field induction.

In magnetostatic vector potential can be written as

$$\vec{A} = \frac{\mu_0}{4\pi} \int\limits_{V} \frac{\vec{J}\,dV}{r} \quad \text{Joules/Amp. or } JA^{-1}$$

here V is the volume of the source surface.

As biot-Savart law in terms of vector potential is very simple hence vector potential $\left(\vec{A}\right)$ can easily be calculated and then magnetic field induction $\left(\vec{B}\right)$ also can be found by taking curl of $\vec{A}$.

Ampere's circuital law

According to ampere's circuital law the line integral of magnetic field $\vec{B}$ around any closed curve is equal to μ_0 times the net current i passing through the area enclosed by the closed curve.

i.e. $\qquad\qquad \oint \vec{B} \cdot \overline{dl} = \mu_0 i$

where μ_0 is free space permeability.

Proof : Consider AB as along, straight conductor with current i, as shown

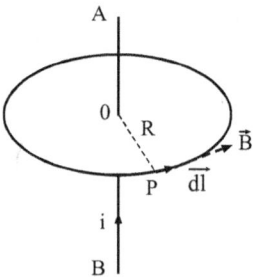

according to Biot-Savart law, magnetic field at P,

$$B = \frac{\mu_0}{2\pi} \frac{i}{R} \qquad \text{...(i)}$$

and at P line integral

$$\oint \vec{B} \cdot \vec{dl} = \oint B\, dl = B \oint dl \qquad \text{...(ii)}$$

Using (i) and $\oint dl = 2\pi R$ in (ii)

$$\oint \vec{B} \cdot \vec{dl} = \frac{\mu_0 i}{2\pi R} (2\pi R)$$

$$= \mu_0 i$$

which is Ampere's circuital law.

This is the integral form of Ampere's circuital law.

Conversion to Differential form

As enclosed current I can be stated as

$$I = \oint_s \vec{J} \cdot \vec{ds} \qquad \text{...(i)}$$

where $\vec{J}$ is the current density and $\vec{ds}$ is the small surfaces area of closed path.

From Ampere's circuital law

$$\oint \vec{B} \cdot \vec{dl} = \mu_0 I$$

$$= \mu_0 \oint_s \vec{J} \cdot \vec{ds} \qquad \text{...(ii)}$$

Stoke's law states that

$$\oint_s (\vec{\nabla} \times \vec{B}) \cdot \vec{ds} = \oint \vec{B} \cdot \vec{dl} \qquad \text{...(iii)}$$

from (ii) and (iii)

$$\oint_s (\vec{\nabla} \times \vec{B}) \cdot \vec{ds} = \mu_0 \oint \vec{J} \cdot \vec{ds}$$

hence

$$\boxed{\vec{\nabla} \times \vec{B} = \mu_0 \vec{J}} \qquad \text{...(iv)}$$

equation (iv) is the differential form of Amperes law. Because $\vec{\nabla} \times \vec{B} \neq 0$, magnetic field is not conservative and its curl has some value.

When the points are inside a closed loop for which $\vec{J} = 0$, $\vec{\nabla} \times \vec{B} = 0$.

APPLICATIONS OF AMPERE'S CIRCUITAL LAW

(i)   Magnetic field induction due to a current carrying straight conductor

Consider a point P at a distance R from the straight conductor.

By symmetry all points at distance r will be on a circle of radius R.

Using Ampere's circuital law for P

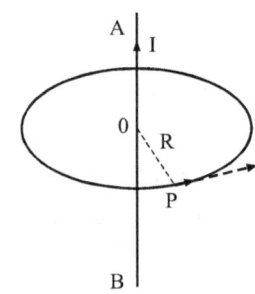

$$\oint \vec{B} . \overrightarrow{dl} = \mu_0 I$$

$$B \oint dl = \mu_0 I$$

$$B \, 2\pi R = \mu_0 I$$

$$\Rightarrow \qquad \boxed{B = \frac{\mu_0 I}{2\pi R}}$$

(ii)   Magnetic field inside an infinitely long current carrying solenoid

If there is a long solenoid of length l as shown.

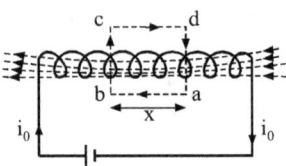

Experimentally it has been observed that magnetic field outside is very small compared with the field inside.

Applying Ampere's circuital law

$$\oint \vec{B} . \overrightarrow{dl} = \mu_0 i$$

$$\oint \vec{B} . \overrightarrow{dl} = \int_a^b \vec{B} . \overrightarrow{dl} + \int_b^c \vec{B} . \overrightarrow{dl} + \int_c^d \vec{B} . \overrightarrow{dl} + \int_d^a \vec{B} . \overrightarrow{dl}$$

here $$\int_b^c \vec{B}.\overline{dl} = \int_d^a \vec{B}.\overline{dl} = 0$$

as both are perpendicular to field lines.

and $$\int_c^d \vec{B}.\overline{dl} = 0 \text{ as } B = 0 \text{ outside the solenoid.}$$

so $$\oint \vec{B}.\overline{dl} = \int_a^b \vec{B}.\overline{dl} = B\int_a^b dl = Bx \qquad ...(i)$$

if there are n turns per unit length with i_0 current in each turn then within length of x, net current enclosed

$$i = nxi_0 \qquad ...(ii)$$

form (i) & (ii)

$$Bx = \mu_0 nx\, i_0$$

or $$\boxed{B = \mu_0 n i_0} \qquad ...(iii)$$

if the solenoid is wrapped on a core of permeability μ_m then

$$B = \mu_m n\, i_0 = \mu_0\, \mu_r n i_0 \qquad ...(iv)$$

(iii) **Magnetic field of induction due to a current carrying cylinder**

Consider a cylinder of radius R with current I passing through it.

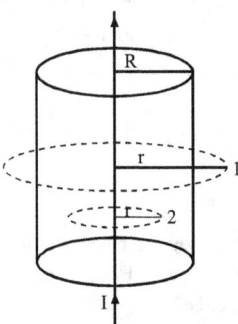

Magnetic field at any point at r distance form the axis of cylinder where (r > R) will be

$$\oint \vec{B}.\overline{dl} = B \oint dl = B2\pi r \qquad ...(i)$$

as r > R current will pass through the circuit of radius r.

so $B_2 \pi r = \mu_0 I$

$$\boxed{B = \frac{\mu_0 I}{2\pi r}}$$ ...(ii)

It is same as for an infinite line current element.

When r < R

Consider a Gaussian surface of radius r inside the cylinder and current enclosed by the inner circle of radius r is given by

$$I' = \frac{Current}{Area\ of\ the\ cylinder} \times area\ of\ the\ inner\ circle$$

$$I' = \frac{I}{\pi R^2} \times \pi r^2 = \frac{I r^2}{R^2}$$ ...(iii)

so $B2\pi r = \mu_0 I' = \mu_0 \dfrac{I r^2}{R^2}$

so $\boxed{B = \dfrac{\mu_0 I r}{2\pi R^2}}$ for r < R   ...(iv)

Inside a hollow cylinder

When there is a hollow cylinder, then whole current will exist only on the surface of the cylinder and inside current will become zero, so I = 0

hence $B\ 2\pi r = 0$

or $\boxed{B = 0}$

Time varying fields

When any charge is at rest it produces electrostatic fields and when charge moves with constant velocity, magnetic field is produces . Here electric and magnetic fields are independent of each other in the time invariant fields.

Whenever any charge moves with an acceleration, it produces a time varying electric field which then produces changing magnetic field. In the same manner a time varying magnetic field produce a changing electric field. Hence oscillating electric field produces an oscillating magnetic field and oscillating magnetic field produces oscillating electric field resulting in a wave like electric and magnetic fields known as electromagnetic wave, which can transport energy over a very long distances. Michael Faraday proposed the basis of electromagnetic field concept and gave postulates for electromagnetic induction to relate time varying fields.

Electromagnetic Induction

Faraday in 1831, found that when there is a change in magnetic flux passing through any circuit, an e.m.f. is generated in the circuit. A current flows in the circuit known as induced current and emf as induced emf. This lasts till the change in flux exists. As soon as the change stops, no emf or current is induced. This whole phenomenon is called electromagnetic induction.

Faraday's Laws of Electromagnetic Induction

1st **Law :** When number of magnetic lines of force attached to a closed circuit in changed, an emf is induced in the coil which lasts up to the change of field lines.

2nd law : The magnitude of induced emf is directly proportional to the rate of change of magnetic flux associated with the coil.

If $\Delta\phi$ is the change in magnetic flux in Δt time interval than the induced emf.

$$e = \frac{\Delta\phi}{\Delta t}$$

as $\Delta t \to 0$

$$e = \lim_{\Delta t \to 0} \frac{\Delta\phi}{\Delta t}$$

$$\boxed{e = \frac{d\phi}{dt}} \text{ volts}$$

if there are N turns in the coil, then the emf induced in the complete coil

$$e = N\frac{d\phi}{dt} \text{ or } \frac{d(N\phi)}{dt}$$

here $N\phi$ is called number of magnetic flux linkages. Its unit is weber turns.

3rd **law :** The direction of induced emf is such that it always opposes its cause of generation. It is called Lenz's law.

Hence

$$e = -N\frac{d\phi}{dt}$$

accordingly direction of induced emf can be found using flemming right hand rule.

Integral Forms

From Faraday's law of electromagnetic induction

$$e = -\frac{d\phi}{dt} \qquad \text{...(i)}$$

and magnetic flux ϕ through a closed surface is

$$\phi = \oint_s \vec{B} \cdot \vec{ds} \qquad \text{...(2)}$$

If the electric field induced is $\vec{E}$, then induced emf around the boundaries of closed surface is

$$e = \oint_c \vec{E} \cdot \vec{dl} \qquad \text{...(3)}$$

from (i), (ii) and (iii)

$$\oint_c \vec{E} \cdot \vec{dl} = -\frac{d}{dt} \oint_s \vec{B} \cdot \vec{ds}$$

or

$$\boxed{\oint_c \vec{E} \cdot \vec{dl} = -\oint_s \frac{\partial \vec{B}}{\partial t} \cdot \vec{ds}} \qquad \text{...(4)}$$

The equation (4) is known as the integral form of Faraday law of electromagnetic induction.

Differential form

From stoke's law we know that

$$\oint_c \vec{E} \cdot \vec{dl} = \oint_s \left(\vec{\nabla} \times \vec{E} \right) \cdot \vec{ds} \qquad \text{...(5)}$$

From (4) and (5)

$$\oint_s \left(\vec{\nabla} \times \vec{E} \right) \cdot \vec{ds} = -\oint_s \frac{\partial \vec{B}}{\partial t} \cdot \vec{ds}$$

or

$$\oint_s \left[\left(\vec{\nabla} \times \vec{E} \right) + \frac{\partial \vec{B}}{\partial t} \right] \cdot \vec{ds} = 0$$

or

$$\boxed{\vec{\nabla} \times \vec{E} = -\frac{\partial \vec{B}}{\partial t}} \qquad \text{...(6)}$$

Equation (6) represents the differential form of Faraday's law of electromagnetic induction.

SOLVED EXAMPLES

Q.1 : A parallel plate capacitor has a separation of 4 mm and potential difference of 200 volt between its plates. The capacitor is placed in a uniform magnetic field B . An electron projected vertically upward parallel to the plates with a velocity of 10^6 m/s, passes between the plates undeflected. Find the magnitude and direction of the magnetic field B between the plates.   [IIT 81]

Sol.:

```
     A              B
  +  |           |  -
  +  |           |  -
  +  |           |  -
  +  |   →       |  -
  +  |   B       |  -
  +  |           |  -
  +  |           |  -
  +  |           |  -
  +  | e⁻ ↑⊖     |  -
  +  |           |  -
```

Suppose V is the potential difference between two plates A and B then

$$E = \frac{V}{d} = \frac{200V}{4 \times 10^{-3} m} = 5 \times 10^4 \text{ V/m}$$

and force due to this electric field on e⁻

$$F = eE \qquad \qquad \qquad ...(i)$$

and force on the e⁻ due to presence of magnetic field

$$F = evB \qquad \qquad \qquad ...(ii)$$

so from (i) and (ii)

$$eE = evB$$

$$B = \frac{E}{v} = \frac{5 \times 10^4}{10^6} = 5 \times 10^{-2} \text{wb/m}^2$$

Q.2 : An electrons is moving in a magnetic field of intensity 10^{-2} wb/m² with a velocity of 10^7 m/s in a circular path of radius 0.57 cm. Find out the specific charge of electron.

Sol.: When particle moves in circular path, necessary centripetal force $\dfrac{\left(mv^2\right)}{r}$ is maintained with lorentz force due to magnetic field i.e. qvB.

So for electron

$$evB = \frac{mv^2}{r} \Rightarrow \frac{e}{m} = \frac{v}{Br}$$

$$\therefore \qquad \frac{e}{m} = \frac{10^7}{10^{-2} \times \left(0.57 \times 10^{-2}\right)} = 1.76 \times 10^{11} \text{ C/kg}$$

Q.3 : An electron after being accelerated through a P. D. of 10^4 V enters a uniform magnetic field of 0.04 T perpendicular to its direction of motion. Calculate the radius of curvature of its trajectory.

Sol.: When electron is accelerated with potential V then

$$\frac{1}{2}mv^2 = eV \qquad \qquad \qquad ...(i)$$

and in magnetic field B, electron takes circular path and centripetal force

$$\frac{mv^2}{r} = evB \qquad \qquad \qquad ...(ii)$$

from (i) and (ii)

$$r = \frac{1}{B}\sqrt{\frac{2mV}{e}}$$

$$= \frac{1}{0.04}\sqrt{\frac{2 \times 9.1 \times 10^{-31} \times 10^4}{1.6 \times 10^{-19}}}$$

$$= 84.3 \times 10^{-4} \text{ m}$$

$$= 8.43 \text{ mm}$$

Q.4 : In cylindrical coordinates, $B = \left(\dfrac{4}{r}\right)\hat{\phi}$ T. Determine the magnetic flux ϕ crossing the plane surface given by $0.5 \le r \le 2.5$m and $0 \le z \le 2.0$m.

Sol.: We know that flux

$$\phi = \oint \vec{B} . \overrightarrow{ds}$$

$$= \int_0^2 \int_{0.5}^{2.5} \left(\frac{4}{r}\right) \hat{a}_\rho \text{ dr dz } \hat{\phi}$$

$$= 8 \ln \frac{2.5}{0.5} = 12.88 \text{ wb}$$

Q.5 : What is the magnitude of force on a wire of length 0.02 m placed inside a solenoid near its centre making an angle of 30° with its axis? The wire carries a current of 6A and the magnetic field due to solenoid is 0-25 T.

Sol.: The force on any conductor of length l is given by

$$F = I\,l\,B\sin\theta$$

$$= 6 \times 0.02 \times 0.25 \times 0.5$$

$$= 0.015 \text{ N}.$$

Q.6 : A horizontal overhead power lines carries a current of 90A from East to West. Compute the magnetic field generated at a distance of 1.5 m below the line.

Sol.: The magnitude of magnetic field due to a long, straight conductor is given by

$$B = \frac{\mu_0 I}{2\pi R}$$

here $\quad\quad\quad \dfrac{\mu_0}{2\pi} = 2 \times 10^{-7} \text{ N/A}^2$

$$= \left(2 \times 10^{-7}\right) \times \frac{90}{1.5}$$

$$= 1.2 \times 10^{-5} \text{ NA}^{-1} \text{ m}^{-1}$$

Q.7 : A differential current element with 10 Amp current and length 2×10^{-3} m is located at (2, 0, 0). Calculate the magnetic field $\vec{B}$ due to this element at (0, 0, 2).

Sol.: Here

$$I = 10 \text{ amp}$$

$$dl = 2 \times 10^{-3} \text{ m}$$

and $\quad\quad\quad \vec{dl} = 2 \times 10^{-3}\,\hat{i} \text{ m}$

so $\quad\quad\quad I \cdot dl = 2 \times 10^{-2}\,\hat{i}$

the distance between point and current element

$$\vec{r} = (0-2)\hat{i} + (0-0)\hat{j} + (2-0)\hat{k}$$

$$= -2\hat{i} + 2\hat{k}$$

$$|\vec{r}| = \sqrt{2^2 + 2^2} = \sqrt{8} = 2\sqrt{2}$$

and
$$\hat{r} = \frac{-2\hat{i} + 2\hat{k}}{\sqrt{8}} = \frac{2\left(-\hat{i} + \hat{k}\right)}{2\sqrt{2}}$$

$$= \frac{-\hat{i} + \hat{k}}{\sqrt{2}}$$

from Biot Savart law

$$\vec{B} = \frac{\mu_0}{4\pi} \frac{I\,\overrightarrow{dl} \times \hat{r}}{r^2}$$

$$= \frac{4\pi \times 10^{-7}}{4\pi} \times \frac{0.02\,\hat{i}}{\left(\sqrt{2}\right)^2} \times \frac{2\left(-\hat{i} + \hat{k}\right)}{\sqrt{2}}$$

$$= -\frac{0.02 \times 10^{-7}}{\sqrt{2}}\,\hat{j}$$

$$= -\frac{2 \times 10^{-9}}{\sqrt{2}}\,\hat{j} = -\sqrt{2} \times 10^{-9}\,\hat{j}$$

$$= -1.414 \times 10^{-9}\,\hat{j} \quad \text{weber/m}$$

Q.8 : A helium nucleus is completing one round of a circle of radius 0.8 m in 4 seconds. Show that the magnetic field at the centre of the circle is 0.5×10^{-19} μ_0 T.

Sol.: As we know that helium nucleus has a charge of $+ 2e$ hence it is considered as a circle of radius r meter equivalent to a current loop and the centre, the magnetic field

$$B = \frac{\mu_0 I}{2r}\ T$$

if one revolution takes t sec then the current is

$$I = \frac{2e}{t}\ \text{amp}$$

so
$$B = \frac{\mu_0 2e}{2rt} = \frac{\mu_0 e}{rt}$$

$$= \frac{\mu_0 \times 1.6 \times 10^{-19}}{0.8 \times 4}$$

$$= 0.5 \times 10^{-19}\ \mu_0\ T\ .$$

Q.9 : The magnetic flux threading a coil changes form 12×10^{-3} wb to 6×10^{-3} wb in 0.01 s. Calculate the induced emf.

Sol.: According to Faraday's law, the induced emf is

$$e = -\frac{\Delta(N \phi)}{\Delta t}$$

$$= -\frac{\left(6 \times 10^{-3}\right) - \left(12 \times 10^{-3}\right)}{0.01}$$

$$= 0.6 \text{ wb/s}$$

$$= 0.6 \text{ V}$$

Q.10 : A coil having 100 turns and area 0.20 m² is placed normally in a magnetic field . the field changes from 0.20 wb/m² to 0.60 wb/m² uniformly over a period of 0.01 s. Calculate the emf induced in the coil.

Sol.: For each coil placed perpendicular to a magnetic field B is given by

$$\phi = BA$$

and change in flux due to change in B is

$$\Delta \phi = (\Delta B) A$$

$$= (0.60 - 0.20) \text{ wb} / m^2 \times 0.20 m^2$$

$$= 0.08 \text{ wb}$$

By Faraday law, the magnitude of the induced emf is

$$|e| = N \frac{\Delta \phi}{\Delta t}$$

$$= \frac{100 \times 0.08}{0.01}$$

$$= 800 \text{ V}$$

Q.11 : When a magnetic flux lines changes from 5.5 × 10⁻⁴ to 5 × 10⁻⁵ in 0.1 sec through a coil of resistance 10 ohm with 1000 times. Find the electromotive force and the charge flowing through the coil.

Sol.: Here change in flux

$$\Delta \phi = \left(5 \times 10^{-5}\right) - \left(5.5 \times 10^{-4}\right)$$

$$= -50 \times 10^{-5} \text{ wb}$$

hence induced emf

$$e = -N\frac{\Delta\phi}{\Delta t}$$

$$= -1000 \times \frac{\left(-50 \times 10^{-5}\right)}{0.1}$$

$$= 5V$$

the developed current $I = \dfrac{e}{R}$

$$= \frac{5V}{10\Omega} = 0.5 \text{ Amp}.$$

so the charge passed through the coil is

$$q = I \times \Delta t = 0.5 \times 0.1 = 0.05 \text{ C}$$

Q.12 : A current distribution gives rise to the vector potential $\vec{A} = x^2 y\hat{i} + y^2 \times \hat{j} - 4xyz\,\hat{k}$ **wb/m. Calculate (i)** $\vec{B}$ **at (–1, 2, 5) (ii) magnetic flux through the surfaces defined by z = 1,** $0 \le x \le 1,\ -1 \le y \le 4$. **[RU 2001]**

Sol.: As curl of the vector potential will give the magnetic field intensity

$$B\,(-1,\,2,\,5) = \text{curl } \vec{A}$$

$$= \begin{vmatrix} \hat{i} & \hat{j} & \hat{k} \\ \dfrac{\partial}{\partial x} & \dfrac{\partial}{\partial y} & \dfrac{\partial}{\partial z} \\ x^2 y & y^2 x & -4xyz \end{vmatrix}$$

$$= \left[-4xz\right]\hat{i} - \left[-4yz\right]\hat{j} + \left(y^2 - x^2\right)\hat{k}$$

$$= 20\,\hat{i} + 40\hat{j} + 3\hat{k} \text{ wb/m}^2$$

and flux $\phi = \oint \vec{B}.\,\overrightarrow{ds}$

where $\vec{B} = -4xz\,\hat{i} + 4yz\,\hat{j} + \left(y^2 - x^2\right)k^2$

and $$\overrightarrow{ds} = dx \, dy \, \hat{k}$$

so the flux $$\phi = \oint \vec{B} \cdot \overrightarrow{ds}$$

$$= \int\limits_{-1}^{4} \int\limits_{0}^{1} \left[-4xz\,i + 4yz\,\hat{j} + \left(y^2 - x^2\right)\hat{k} \right] dx \, dy \, \hat{k}$$

$$= \int\limits_{-1}^{4} \int\limits_{0}^{1} \left(y^2 - x^2\right) dx \, dy = \int\limits_{-1}^{4} \left[y^2 \left(-\frac{1}{3}\right) \right] dy$$

$$= \left[\frac{y^3}{3} - \frac{y}{3} \right]_{-1}^{4} = \frac{60}{3} = 20 \text{ wb}$$

Q.13 : Copper has 8×10^{28} free conduction electrons/m³. A copper wire of length 2.0 m and cross sectional area 8×10^{-6} m² carrying a current and lying perpendicular to a magnetic field of 5×10^{-3} T experiences a force of 8×10^{-2} N. Calculate the drift velocity of free electrons in the wire.

Sol.: When free electrons move with velocity v_d in the wire then the current is

$$I = ne \, A \, v_d \qquad \qquad ...(i)$$

and force on the conductor

$$F = I \, Bl \qquad \qquad ...(ii)$$

so $$I = \frac{8 \times 10^{-2} \, N}{\left(5 \times 10^{-3}\right) \times 2m}$$

$$= 8A$$

and $$v_d = \frac{I}{neA} = \frac{8A}{8 \times 10^{28} \times 1.6 \times 10^{-19} \times 8 \times 10^{-6}}$$

$$= 0.78 \times 10^{-4} \text{ m/s}$$

Q.17 : A current of 20 Amp flows in downward direction in a long straight vertical wire and magnetic flux density in horizontal direction is 2×10^{-5} T. what is the distance of the neutral point form the wire?

Sol.: Neutral point will be where horizontal magnetic field will become equal to the magnetic field produced by current carrying conductor.

So
$$B_H = \frac{\mu_0 I}{2\pi r}$$

$$2 \times 10^{-5} = \frac{4\pi \times 10^{-7} \times 20}{2\pi \times r}$$

$$r = 0.2 \text{ m}$$

Q.18 : A circular coil is placed in uniform magnetic field of 0.10 T normal to the plane of the coil. If the current is 5.0 A in the coil, Find (a) total torque on the coil (b) total force on the coil (c) average force on each electron due to magnitude field (The coil in made of copper wire of cross section 10^{-5} m² and free electron density in copper is 10^{29} m⁻³).

Sol.: We know that the torque on a current carrying coil of area A is

$$\tau = NIAB \sin\theta$$

(a)   here $\theta = 0$ i.e. $\sin\theta = 0$ so $\tau = 0$

(b)   the net forces will always zero.

(c)   and magnitude of force on a free electron

$$F = ev_d B$$

$$= \frac{IB}{nA} \qquad\qquad (\text{as} \quad I = neAv_d)$$

$$= \frac{5 \times 0.10}{10^{29} \times 10^{-5}} = 5 \times 10^{-25} \text{ N}$$

Q.19 : In the figure two current carrying wires are A and B. Find the magnitude and directions of the magnetic field at points 1, 2 and 3.

Sol.:

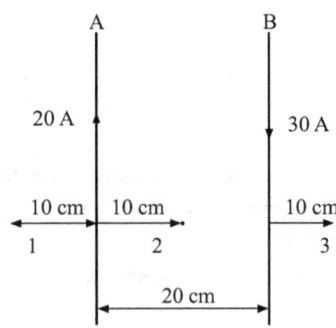

We know that magnetic field at R distance form straight current carrying wire

$$B = \frac{\mu_0 I}{2\pi R}$$

$$= \left(2\times10^{-7}\right)\frac{I}{R} \text{ NA}^{-1}\text{ m}^{-1}$$

Due to current in wire A at point 1 field will perpendicular upward and at 2 and 3 downwards and similarly due to current in wire B, 1 and 2 will be downward and 3 will be upward.

Hence points 1 and 3 are in opposite fields whereas point 3 will be always in same direction.

So resultant field at 1

$$B = B_1 - B_2$$

$$= \left(2\times10^{-7}\right)\frac{20}{0.1} - \left(2\times10^{-7}\right)\times\frac{30}{0.3}$$

$$= 2\times10^{-5} \text{ NA}^{-1}\text{m}^{-1} \text{ perpendicular and upward to page.}$$

at 2

$$B = B_1 + B_2$$

$$= \left(2\times10^{-7}\right)\left[\frac{20}{0.1} + \frac{30}{0.3}\right]$$

$$= 1\times10^{-4} \text{ NA}^{-1}\text{m}^{-1} \text{ perpendicular and downward to page.}$$

at 3

$$B = B_2 - B_1$$

$$= \left(2\times10^{-7}\right)\left[\frac{30}{0.1} - \frac{20}{0.3}\right]$$

$$= 4.7\times10^{-5} \text{ NA}^{-1}\text{m}^{-1} \text{ perpendicular and upward to the page.}$$

Q.20 : 8.0 cm length of a conductor is placed parallel to 2m length of a conductor at a distance of 2.0 cm. The conductors carry currents of 2 and 5 Amp respectively in opposite direction. Find the total force exerted on the long conductor.

Sol.: We know the magnetic field near to any straight conductor

$$B = \frac{\mu_0 I_1}{2\pi R}$$

So the force experienced by the short conductor carrying a current I_2

$$F = I_2 \, Bl$$

$$= \frac{\mu_0}{2\pi} \frac{I_1 I_2 l}{R}$$

so

$$F = \left(2\times10^{-7}\right) \times \frac{5\times2\times8\times10^{-2}}{2\times10^{-2}}$$

$$= 8\times10^{-6}\,N$$

according to Newton's third law, the long conductor will also experience an equal repulsive force 8×10^{-6} N due to the small conductor.

Q.21 : An air-cored solenoid of length 50 cm and area of cross-section 28 cm² has 200 turns and carries a current of 5A. On switching off, the current decreases to zero within a time interval of 2 ms. Find the average induced emf across the ends of the switch.

Sol.: As we know magnetic field

$$B = \mu_0 n \, I \qquad\qquad (n = \text{No. of turns/length})$$

$$= \left(4\pi\times10^{-7}\right) \times \left(\frac{200}{50\times10^{-2}}\right) \times 5$$

$$= 25\times10^{-4} \, T$$

when switch off, flux reduces to zero, so change in flux.

$$\Delta\phi = 0 - NBA$$

$$= -200 \times \left(25\times10^{-4}\right) \times \left(28\times10^{-4}\right)$$

$$= -14\times10^{-4} \, wb$$

and induced emf

$$e = -\frac{\Delta\phi}{\Delta t} = \frac{14\times10^{-7}}{2\times10^{-3}} = 0.7 \, V$$

SUMMARY

- When any electric charge moves in a combined effect of electric and magnetic field then total force experienced by the charge

$$\vec{F} = q\left[\vec{E} + \vec{v} \times \vec{B}\right]$$

- If a current carrying conductor of length is placed in any magnetic field then the force experienced by that length of the conductor

$$F = I\,\vec{l} \times \vec{B}$$

$$= I\,lb\,\sin\theta$$

- Magnetic flux

$$\phi_B = \oint_s \vec{B}.\,\vec{ds}$$

- Gauss law of magnetism

$$\oint_s \vec{B}.\,\vec{ds} = 0$$

- Divergence of magnetic field is given as

$$\vec{\nabla}.\vec{B} = 0$$

which shows that magnetic poles exist in pairs.

- According to Biot-Savart law, the magnetic field at a point near to any current carrying conductor is given by

$$d\vec{B} = \frac{\mu_0}{4\pi}\frac{I\,\vec{dl} \times \hat{r}}{r^2}$$

- Magnetic field at the centre of a current carrying loop of radius r.

$$B = \frac{\mu_0 I}{2r}$$

- Magnetic field at a point due to current carrying straight conductor

$$B = \frac{\mu_0}{4\pi}\frac{I}{R}\left(\sin\phi_1 + \sin\phi_2\right)$$

$$B = \frac{\mu_0 I}{2\pi R} \; NA^{-1}m^{-1}$$

- Force between two parallel current carrying conductors

$$F = \frac{\mu_0}{2\pi} \frac{I_1 I_2}{R} l$$

this is attractive when currents are in same direction and repulsive when currents are in opposite direction.

- Magnetic field at the axis of current carrying circular coil

$$B = \frac{\mu_0 \; NI \; a^2}{2\left(a^2 + x^2\right)^{3/2}} \; NA^{-1}m^{-1}$$

- Magnetic field of a solenoid

$$B = \mu_0 nI \quad \text{at the centre}$$

$$= \frac{\mu_0 nI}{2} \quad \text{at the ends.}$$

- According to Ampere's circuital law the line integral of magnetic field due to a closed current carrying loop

$$\oint_c \vec{B}.\,\vec{dl} = \mu_0 I \quad \text{Integral form}$$

$$\vec{\nabla} \times \vec{B} = \mu_0 \vec{J} \quad \text{Differential form}$$

- Magnetic field due to a current carrying cylinder

$$B = \frac{\mu_0 I r}{2\pi R^2} \quad \text{for } r < R$$

$$= \frac{\mu_0 I}{2\pi r} \quad \text{for } r > R$$

$$= 0 \quad \text{for a hollow cylinder and } r < R.$$

- According to 1st law of electromagnetic induction, when ever there is a change in magnetic flux linked with any closed solenoid then there is emf produced known as induced emf.

- According to 2nd law of electromagnetic induction the induced emf is given by

$$e = \frac{\Delta\phi}{\Delta t}$$

- According to 3rd law of EMI, the direction of induced emf is such as to oppose its cause of production so

$$e = -\frac{\Delta\phi}{\Delta t}, \text{ this is known as Lenz's law.}$$

- $$\oint_c \vec{E}.\,\overrightarrow{dl} = -\oint_s \frac{\partial \vec{B}}{\partial t}.\,\overrightarrow{ds}\text{ integral form of Faraday electromagnetic law.}$$

- $$\vec{\nabla} \times \vec{E} = -\frac{\partial \vec{B}}{\partial t}\text{ Differential form of Faraday's electromagnetic law.}$$

- **Right hand palm rule No. 2**

  If we stretch palm of right hand such that the stretched fingers points towards the magnetic field direction and thumb points to the direction of current the force on the conductor will be perpendicular to the palm.

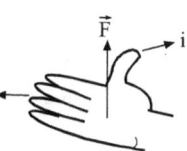

- **Flemming's left hand rule –**

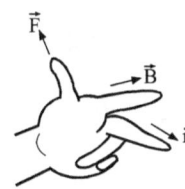

  If the thumb, middle finger and fore finger of the left hand are stretched in mutually perpendicular direction such that fore finger points in the direction of magnetic field $\vec{B}$ and middle finger towards i the current direction then thumb will point in force $\vec{P}$ direction on the conductor.

- **Right hand palm rule no. 1**

  If we stretch right hand palm such that thumb points in current direction and fingers towards the P point at which magnetic field direction is to be found, then the

  perpendicular to the palm will show the direction of $\vec{B}$ at P.

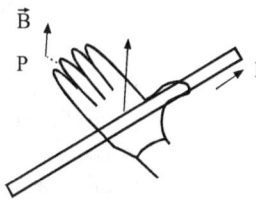

- **Right hand thumb rule –**

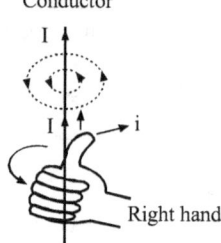
Conductor

Right hand

It is used to find the direction of magnetic field due to a straight current carrying conductor. If we hold the straight conductor in our right hand in such away that thumb points in current direction, then the encircled fingers represents the magnetic field lines around the conductor.

- **Maxwell's right hand screw rule**

Current I

Right hand

If we takes the screw driver in our right hand such that it points to current direction and if we rotates it in such a manner that screw moves in the direction of current, then the direction of rotation of screw driver will tell the direction of magnetic field lines.

- **Direction of Induced current - Fleming's right hand rule–**

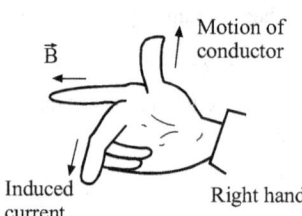
$\vec{B}$ Motion of conductor

Induced current   Right hand

If we stretch the right hand thumb, fore finger and middle finger in mutually perpendicular direction in such a manner that fore finger is towards direction of magnetic field and thumb points in the direction of motion of conductor, then the middle finger will point in the direction of induced current.

EXERCISE

1. Describe the magnitude and direction of force acting on a charge moving in a magnetic field. When it is minimum and when it is maximum?

2. Derive an expression for the force experienced by a current-carrying straight conductor placed in a uniform magnetic field. State the rule to find the direction of this force.

3. Write Biot-Savart law for the magnetic field due to a current element, explaining the symbols.

4. Discuss analogies and differences between coulomb's law and Biot-Savart law.

5. A current is flowing through a thin, straight metallic conductor of infinite length. Find expression for the magnetic field at a distance from it.

6. Derive the relation for the force per unit length between two infinitely-long, parallel, straight conductors carrying current. Hence define one ampere.

7. Derive an expression for the magnetic field at a point on the axis of a circular coil carrying current, and hence at the centre of the coil.

8. Derive the expression for the magnetic field at the centre of a circular current carrying coil.

9. Deduce the expression for the magnetic field produced at the centre of a semicircular wire loop of radius R, carrying a current I.

10. Describe the magnetic field within a long, current carrying solenoid. Obtain expressions for the field within and at the ends of the solenoid.

11. State and explain Ampere's circuited law. Hence derive an expression for the magnetic field due to a solenoid.

12. Calculate, using Ampere's circuital law, the magnetic field due to infinitely-long current carrying conductor.

13. State and explain Faraday's laws of electromagnetic induction.

14. Show that lenz's law follows the principle of conservation of energy.

15. Give integral form of Faraday's laws of electromagnetic induction and convert these into differential forms.

16. Define magnetic flux. Write SI units of magnetic flux and magnetic flux density.

17. Write the formula for the force on a charge q moving with velocity $\vec{v}$ in a uniform magnetic field $\vec{B}$. What is the magnitude of this force? When this will be zero?

18.   Prove that $\vec{\nabla}.\vec{B} = 0$ where $\vec{B}$ is the magnetic flux density.

19.   What is vector potential? How the vector field can be calculated?

20.   Prove that $\vec{\nabla} \times \vec{B} = \mu_0 \vec{J}$.

21.   Deduce the differential form of Ampere's circuital law and hence prove that static magnetic field is not conservative.

22.   Define electromagnetic wave. Can accelerated charged particle produce electromagnetic wave? give reasons.

23.   An electron is moving vertically upward with a speed of 2×10^8 m/s. What will be the magnitude and direction of the force on the electron exerted by a horizontal magnetic field of 0.50 wb/m² directed towards west? What will be the acceleration of the electron? $[1.6 \times 10^{-11} N$ north, 1.8×10^{19} m/s²]

24.   An electron moving with velocity 5×10^7 m/s enters a magnetic field of 1.0 wb/m² at an angle of 30 to the field. Calculate the force on the electron.

                                                              $[4 \times 10^{-12}$ N]

25.   A 2-MeV proton is moving perpendicular to a uniform magnetic field of 2.5 T. Find the force on the proton. [Given mass of proton = 1.65×10^{-27} kg]

                                                       [Ans.: 7.88×10^{-12} N]

26.   A 40 cm long wire carrying a current of 2.5A is placed perpendicular to a magnetic field of 8×10^{-3} wb/m². Find the force experienced by the wire. $[8 \times 10^{-3}$ N]

27.   A current of 5.0 A is flowing upward in a long vertical wire placed in a uniform horizontal north-ward magnetic field of 0.0207. How much forces and in what direction will the field exert on 0.06 m length of the wire. $[6 \times 10^{-3}$ N, west]

28.   A straights wire carries a current of 3A. Calculate the magnitude of the magnetic field at a point 10 cm away from the wire. Draw a diagram to show the direction of the magnetic field. $[6 \times 10^{-6}$ T]

29.   A circular loop of radius 5 cm carries a current of 0.5 amp. Calculate the magnitude of the magnetic field at its centre. $[6.28 \times 10^{-6}$ N/A-m]

30.   Calculate the force per unit length on a long straight wire carrying a current 4 amp due to a parallel wire carrying a current of 6 A. The distance between the wires is 3.0 cm. $[1.6 \times 10^{-4}$ N/m]

31.   Two parallel wires, each of length 2 m and carrying a current of 0.40 A in the same direction, are placed 0.40 m apart in air. Find the force per unit length on each wire.

                                                   $[8 \times 10^{-8}$ N/m attractive]

32. An Air-solenoid has 500 turn of wire in its 40 cm length. If the current in the wire be 1.0 A, find the magnetic field at the axis inside the solenoid.

$$[1.57 \times 10^{-3} \text{ NA}^{-1} \text{ m}^{-1}]$$

33. The magnetic field at the centre of a 50 cm long solenoid is 4×10^{-2} N/(A-m) when a current of 8 amp flows through it. What is the number of turns in the solenoid.

[1990]

34. A 0.5 m long solenoid has 500 turns and has a flux density of 2.52×10^{-3} T at its centre. Find the current in the solenoid.   [2.0 Amp.]

35. A test charge having charge 0.4 C is moving with a velocity of $\left(4\hat{i} - \hat{j} + 2\hat{k}\right)$ m/s through an electric field of intensity $10\hat{i} + 10\hat{k}$ and a magnetic field $2\hat{i} - 6\hat{j} - 6\hat{k}$. Determine the magnitude and direction of the lorentz force acting on the test charge.

[WBUT 2007]

36. If the vector potential $A = \left(x^2 + y^2 - z^2\right)\hat{j}$ at $\left(x_1 y_1 z\right)$. Find the magnetic field at (1, 1, 1).   [WBUT 2007]